Gadamer and Practical Philosophy

ÆR

American Academy of Religion
Studies in Religion

Editor
Lawrence S. Cunningham

Number 64
GADAMER AND PRACTICAL PHILOSOPHY

by
Matthew Foster

GADAMER AND PRACTICAL PHILOSOPHY
The Hermeneutics of Moral Confidence

by
Matthew Foster

Scholars Press
Atlanta, Georgia

GADAMER AND PRACTICAL PHILOSOPHY

by
Matthew Foster

© 1991
The American Academy of Religion

Truth and Method, Second Revised Edition by Hans Georg Gadamer, Translation Revised by Joel Weinsheimer and Donald G. Marshall. Second, revised edition © 1989 by The Crossroad Publishing Company. Reprinted by permission of the publisher.

Reason in the Age of Science by Hans Georg Gadamer © 1981 by the Massachusetts Institute of Technology. Reprinted by permission of the publisher.

Library of Congress Cataloging in Publication Data

Foster, Matthew Robert, 1951-
   Gadamer and practical philosophy : the hermeneutics of moral
   confidence / by Matthew Foster.
   p.   cm. — (Studies in religion / American Academy of Religion
   : no. 64)
   Includes bibliographical references.
   ISBN 1-55540-610-6. — ISBN 1-555540-611-4 (pbk.)
   1. Gadamer, Hans Georg, 1900-   . 2. Ethics, Modern—20th
century. 3. Hermeneutics—History—20th century. I. Title.
II. Series: AAR studies in religion ; no. 64.
B3248.G34F67 1991
170'.92—dc20   91-21396
                                                    CIP

Printed in the United States of America
on acid-free paper

Contents

Acknowledgements

I am grateful to all who helped bring this project to completion. This includes Dr. Susan Thistlethwaite of Chicago Theological Seminary, who found a venue for this text, and Dr. Lawrence Cunningham of the University of Notre Dame, Editor of this Studies in Religion Series, who selected it for publication and provided patient assistance. Rob Massie carefully reviewed the text in its final stages and made valuable suggestions.

For permission to quote from Gadamer's magnum opus, *Truth and Method*, in its 2nd (1989) revised English translation, I am grateful to Crossroad Publishing Company.

For permission to quote from Gadamer's 1981 collection of translated essays, entitled *Reason in the Age of Science*, I am grateful to the Massachusetts Institute of Technology Press.

Notes on References

Hans-Georg Gadamer, *Truth and Method = TM*

References appear in the format of "*TM*, 200 [230]," first indicating source pagination from the 2nd (1989) revised English edition. Since this edition is comparatively new, pagination in brackets for the corresponding passage in the 1st (1975) English edition has been added.

Hans-Georg Gadamer, *Reason in the Age of Science = RAS*

The essays in this book were first published separately in German between 1976 and 1979. References omit the respective essay titles and refer only to this collected and translated work because of the substantive coherence and chronological proximity of the essays it contains.

All references identified without author are works by Gadamer.

Introduction

By one measure, the modern era began when we willed a departure from once-certain realities. Awakening to new powers of mind and making, and loosening our bonds to God, heaven and hell, we craved to see what we really were. Since then we have often discovered—or at least felt—that we are bereft of adequate orientation, and sometimes even of the gravity that would establish our spiritual ground. We have become explorers of human possibility, but unable to return home and unsure of where or how to end our journey. We peer ahead at new choices, once unimaginable, approaching us from the horizon—choices about what to obey and what to ignore, what to create and what to destroy, in nature, in artifice, in society, even in our own bodies. Looking in the other direction, familiar guides to what is holy, moral and good seem to recede ever further. An ancient drama is playing out once more: the taste of the new fruit is always more than we bargained for, yet neither would we choose to return to our former state.

A poignant expression of this condition is the erosion of confidence in moral judgment for many in our time. Of course, in every age people worry about moral decisions. And every age reels from the consequences of some audacity of its own will. In our time, though, we also worry about whether there even are such things as moral decisions, or whether ethics is so relative and so contested as to be meaningless. Today, we must decide not only how to live in order to realize the ends of life; we must also decide what those ends are. When so much is unclear, the whole enterprise comes into question. Is there any real and reliable good

1

at which our moral struggles should aim? Can we have confidence that moral reasoning about reaching that end is worthwhile?

This book takes the erosion of moral confidence as a premise about the age, and examines how the work of Hans-Georg Gadamer can help us understand and address this predicament. We will see that the perspective offered by this contemporary German philosopher, which he calls philosophical hermeneutics, contains resources of contemporary bearing and still unexplored merit. Our focus will be on one prominent project within that perspective, the project to rehabilitate the classical notion of practical philosophy. We will pursue questions about whether and how Gadamer's work is appropriate to the contemporary challenge; our interpretation of Gadamer's 'debates' with Leo Strauss and Jürgen Habermas will be particularly important vehicles in this regard. We will also identify and address other problems in Gadamer's work, problems which in fact do limit its ethical relevance—but unnecessarily. In both respects, this work is a constructive defense of the ideas Gadamer bequeaths to us: although fluid and elusive, those ideas warrant the patience required to distill their essence. The significance of Gadamer's hermeneutical perspective for understanding moral theorizing and moral practice goes beyond his attention to the formal features of practical deliberation: it bears witness to truths which seem to be lost on Gadamer's critics and neglected by his supporters, and of which Gadamer himself seems to be not fully aware.

To imagine a bridge, much less to build one, between the realm of moral practice and the theoretical perspective of philosophical hermeneutics may seem strange, even to the novice in these fields. Hermeneutics originally meant theories about, and manuals for, the interpretation of texts, especially texts in foreign languages and about strange peoples and their beliefs, without regard for the merit of the claims made in those texts. And today, although hermeneutics has acquired a larger scope and more philosophical character, the moral or simply normative significance of these inquiries still appears to lie beyond its competence, and perhaps beyond its concern as well. By contrast, moral practice involves judgments about the socially and politically ordered human good and the historical realization of that end; reflection on such vital questions, and especially on whether it is even meaningful to wrestle with them, may seem quite unrelated to hermeneutics. But as a number of Gadamer's works demonstrate, he clearly believes that there are resourceful connections between moral theory and hermeneutical theory, which arise because of the affinity between moral and interpretive acts.

Yet even to those aware of these ideas in Gadamer's thinking, the goal of substantiating such a bridge—especially demonstrating its capacity to support confidence in our own moral judgments—is a goal beset by difficulties. Critics of Gadamer have been dubious about the notion that there is any substantive connection between moral judgment and hermeneutical understanding, and suspicious that Gadamer actually abets the very anxiety about the moral life that we hope he can help dissolve. Gadamer's advocates pose different challenges; they have sometimes been overly impressed with the connections which Gadamer only sketches, and too ready to assume that his bridge, even when finished, will bear all sorts of traffic. There is a reluctance to simply ask, in regard to legitimate expectations of his readers or implied claims of his own, that his philosophy either speak up or define its limits.

The Nature of Practical Judgment

In order to explore these issues, we begin with some assumptions about the character of practical judgment. The central issue of practical judgment is to identify valid and appropriate connections between universals and particulars—that is, between abstract or general moral values or principles, and the concrete, contingent reality which poses practical problems to us. The aim is to draw the conclusion and make the decision by which we intend to guide our action. Since Aristotle, many have called this process the practical syllogism, composed of a major premise (a universal), a minor premise (a particular), and the deduction inferred from these (the judgment on which to base action).

The problem faced in making connections between universals and particulars in practical judgment is ubiquitous: while it is necessary to establish such a relationship because it points the way toward the conclusion, there is never absolute certainty that the correct relationship between them has been discerned (not to mention perennial doubts about the validity of major premisses or the accuracy of minor premisses). The peculiar character of ethical consciousness is evident in this situation: in its uncertainty we see both the freedom and anxiety of ethical existence, and in its imperative character, we find ourselves compelled to identify good and evil, to choose to create or destroy, to assign merit and blame. The only way in which a given relationship of universals and particulars can be affirmed, the only extent to which it can be justified, is, finally, on the basis of a confidence or trust—even a slim one—since certainties of the kind for which we often wish are simply unavailable. But even confidence can be tested, and must be, for this is how we acquire any new confidence in our judgments and develop the practical wisdom on which well-done deliberation depends.

The ultimate premise of all practical confidence is that there is a wholeness to our existence, to the world; a conviction that universals and particulars not only should be joined, as conscience may dictate, but that reality permits them to be joined. A corollary to this premise is that there is some kind of unity between the grounds of confidence and the normative content of the universals we apply in practical judgments. That is, deliberative confidence (i.e., that we can judge wisely) and substantive confidence (i.e., that the content of our decision and our action is good) may involve different questions, but they cannot be ultimately dissimilar or incompatible. If this unity is lost, as often seems the case today, then the process of deliberation and the ends of deliberation can only be affirmed on different and incommensurable grounds; ultimately, practical judgment is distorted and confidence in it atrophies and succumbs to confusion and anxiety.

One basic task of a philosophy of practice, therefore, is to inquire into the grounds for confidence in practical judgments. What kind and level of confidence ought we to have when we proceed with a given practice? What makes confidence courageous instead of rash, or lack of confidence cautious instead of cowardly? In whose truths or traditions do we actually have confidence? When are false grounds for confidence foolishly embraced? And in what could we have confidence but do not?

Modern Paralyses of Moral Confidence

The problems we face in the real world today derive part of their difficulty from our uncertainty about how to answer such questions—in short, about whether we can have confidence in our deliberations or in the good at which we aim. In fact, substantive problems and procedural problems work together, in collusion as it were, to unceasingly deter, confuse and unbalance our efforts. Consider for a moment some of the major issues we face.

One issue concerns the political values on which democracies like ours build their social life. Some version of democratic liberalism guides the economic and political life of these nations, but it is often evident that our institutions are agnostic about the ultimate basis of such values; by default, those institutions project and defend such values as pluralism, tolerance toward strange beliefs, civilized competition, and uninhibited consumption as ends in themselves. Values like these may serve us well when life is relatively easy. But as presently conceived, they are free-floating values, unable to adequately tell us why, beyond the need for civil peace, we should be tolerant; nor can they explain what the ultimate value of plurality is, or why economic growth is edifying. When push comes to shove, these values may well be poor indicators of the deeper desires that will likely emerge pre-eminent. We must reconsider the ultimate foundations of our political life. Moreover, with communist structures discredited or dissolved in various lands, there is additional and timely significance to this task.

A second issue is technology. Our age has put enormous faith in technical means, and probably still more in technical ends. Even when the by-products of technology have posed unexpected problems, we have believed that the same levels and methods of technology as we already employ can also cure the ills they produce. Perhaps now we are at least aware that the trinity of science, machines and unending growth has been seducing us with promises of better tomorrows and frightening us with threats to withdraw its favors if we do not obey. Under the shadow of such a god, can we really exercise sufficient control over ourselves to choose how to use technology? Or someday will an aroused and implacable Nature make the choice for us, imposing on us the total payment of debts we have perennially deferred? Many elements of such a future prospect are debatable, but it would be prudent to expect that technological civilization is on a course not only of new and good creativity, but also of increasing complexity and fragility, of human disparity and suffering on a Malthusian scale, and of general socio-political transformation. The issue is more than just a technical choice about the least costly and most sustainable course of economic activity: our prospects call into question the very relationship we have had to nature, and to the tools by which we mix ourselves with it.

A third problem of moral confidence concerns religious life. While some religious communities are obviously thriving, many are not, and their dilemma, which is particularly familiar to those in the West, appears throughout the world wherever "modernizing" forces induce doubt about religious approaches to life. The problem is acute for those who want to affirm religious truths, and who simultaneously perceive historical forces, the worldly forces of society, psyche and culture, as primary factors in causality—perhaps not as the only factors, but prominent in every event, even the holiest. They become self-conscious about the beliefs of mythic proportion announced for centuries in their own traditions. The religious meanings which they are still able to embrace are also disembodied and fragmented meanings. Somehow, as a result, the relevance of even this remnant to the questions of life is suspended again and again. Some then try to find a new naivete through cunning revisions of their tradition; others give up and sedate their souls in order to simply function in life. The religious loyalists, who remain within their institutions, cannot escape this problem either, though it affects them differently. Not the least of these manifestations are the taming of the received legacy for the benefit of modern sensibilities; the politicization (in every direction) of what remains; and in some cases declining numbers of laity and clergy. Such developments raise basic questions about the meaning—and viability—of these institutions. For both those inside some religious fold and those outside of any, the atrophy of confidence in ultimate truth and goodness is a practical issue, not only a theoretical or theological one: the health of individuals, and no less of societies, depends on this religious glue. Its adhesive quality can be a nuisance, or even an obstruction. This changes when we become aware of some critical joint in life that would otherwise break apart, when we feel some hidden tether to our origin and destiny unexpectedly going limp. Indeed, without confidence in the ultimate bonds of which religion speaks, it is difficult to see how the other two issues can be wisely addressed.

What is common to these three problems is that they appear to have no solution: they lack a self-evident good which could be chosen from among alternative solutions. Moreover, these three issues almost seem to be predicated on excluding the possibility of a judgment that fits our expectation of what a 'solution' should look like. Consider the inadequacy of some of the available choices in the issues just raised. If plurality is conceivably an end in itself, it is still an empty one today, neutered by reduction to its political utility. But if plurality is instrumental to some other end, the civil function of plurality will be undercut when institutions announce their adherence to that end. If we begin to suspect that every use of technology leaves us with morally dirty hands, can any proposed constitution for technology be trusted? But if we cannot simply absent ourselves from technical life either, does it matter what we choose? If we live by the stories and icons of one religious community, it seems we must do so in a mental chamber foreign to many around us and even to the rest of our own thinking. But if we live first by the myth of harmonious pluralism, before what are we really bowing, and to whom have we entrusted our destiny?

These and other issues of contemporary practical concern must remain largely in the background of this investigation. We will be indirectly concerned with the features of all such issues—morphological, methodological, and ontological—which raise the question of confidence in our moral capacities, and confidence in the reality of the good with which we hope such capacities cohere.

There is, however, one causal factor—an intellectual premise in the background of our practical decisions about all these issues—which draws us nearer the issues of this investigation. This is the notion, born of the modern scientific mind, that there is not simply a difference but an absolute dichotomy and incompatibility between fact and value. The merits of this 'incompatibility thesis' are well-known in various fields of investigation and endeavor; but its fundamental effect on practical judgment is to cast doubt on whether universals and particulars can justifiably be connected—precisely the premise on which practice most depends. The universals which are applied in practical judgment are considered in our scientific ethos to be unverifiable values, while particulars are considered to be verifiable facts; since only what is verified by scientific methods can count as real in this ethos, we frequently see evidence around us of a tacit conclusion that there is no way to link universals and particulars in the way that practical judgment in fact requires.

In short, confidence in the possibility of meaningful moral application of universals to particulars cannot be justified in the light of this thesis. It is no wonder that we often feel divided against ourselves, paralyzed between the necessity and the impossibility of moral judgment. This is perhaps the most important, yet least evident reason why the most widespread moral concern of our time is not any particular issue over all others, but is whether moral reasoning and moral acting have any validity at all. Since wise practical judgment seems impossible, it becomes ever more doubtful that good can result or appear in human action.

This is not to say that the absence of self-evident solutions to some of our practical problems is solely due to this incompatibility thesis; it is obvious that, in our time, we are juggling several problems at once. But it is important to investigate any signs of a connection between the methodological and substantive problems we face; moreover, the basic premise of practical reason, as noted above, implies that there is such a connection. If no solution is apparent, perhaps it is because we have forgotten, in our scientific worldview, what an authentic solution looks like, and therefore what authentic deliberation presumes. And if the division of fact from value makes us unable to know what moral practice means, we must either accept that we are amoral beings, or we must mend this divorce.

Gadamer's Claim

The interconnecting issues just posed will form the background on which we pursue our primary inquiry into how Gadamer's philosophical perspective on human understanding illuminates the nature of moral reasoning in particular, and of the practice to which such reasoning leads. *Truth and Method* lays out the foundation of Gadamer's perspective, and is an important resource for his view of moral judgment as well. It was first published in Germany in 1960; an English translation appeared in 1975,[1] and a significantly improved revision of that translation appeared in 1989 on which the present work will rely for quotation.[2] Gadamer deals more directly with moral practice and moral theory, and with their relationships to hermeneutics, in a number of essays written in the 1970s and 1980s. This is notably true of the eight essays, first published between 1976 and 1979, which appear in the collection entitled, *Reason in the Age of Science.*[3] Passages from a variety of Gadamer's other works in English translation will also be of relevance.[4]

Gadamer's own words provide the best warrant for our inquiry, and also help focus its central goal. Two passages from a 1970 essay, one a question and the other a reply, convey not only a claim for the theoretical relevance of philosophical hermeneutics for moral theory, but also a claim for the significance of its concrete application today:

> How can we learn to recover our natural reason and our moral and political prudence? In other words, how can we reintegrate the tremendous power of

---

[1] Hans-Georg Gadamer, *Truth and Method*, translator unidentified (although the "Translator's Preface" of the 2nd revised edition indicates that it was W. Glen-Doepel), based on the 2nd (1965) German edition, ed. John Cumming and Garrett Barden (New York: Seabury Press, 1975).

[2] *Truth and Method*, 2nd revised edition, translation revised by Joel Weinsheimer and Donald G. Marshall, based on the 5th (1986) German edition (New York: Crossroad Publishers, 1989).

[3] Hans-Georg Gadamer, *Reason in the Age of Science*, trans. Fredrick G. Lawrence (Boston: MIT Press, 1981).

[4] One notable exception to this use of existing translations will be reference to an important essay from 1963 which, surprisingly, is still unpublished in English translation: "Über die Möglichkeit einer philosophischen Ethik." Certain relevant passages of this essay are rendered here in English for the first time.

our technique within a well-balanced order of the society and reconstitute a living solidarity? . . .

. . . Precisely and especially practical and political reason can only be realized and transmitted dialogically. I think, then, that the chief task of philosophy is to justify this way of reason and to defend practical and political reason against the domination of technology based on science. That is the point of philosophical hermeneutic. It corrects the peculiar falsehood of modern consciousness—the idolatry of the sciences; and it vindicates again the noblest task of the citizen—decision-making according to one's own responsibility instead of conceding that task to the expert. In this respect, hermeneutic philosophy is the heir of the older tradition of practical philosophy.[5]

Gadamer asserts, first, that contemporary political reflection and action have atrophied under our application of scientific method, and second, that the purpose of philosophical hermeneutics is ultimately to "defend practical and political reason," and in this regard is heir to Aristotle's philosophy of practice. The first claim is one Gadamer shares with many other writers. The second is an original and even startling claim, since, as suggested at the outset, hermeneutics hardly seems to have an obvious connection to practical reason, or a clear descent from Aristotle, for that matter. These passages could, in fact, be considered an epigram of Gadamer's nascent philosophy of practice. They could equally serve as criteria for testing his effort.

The Possibility of a Hermeneutical Philosophy of Practice

The aim of this study can now be more formally stated. The resources of philosophical hermeneutics will be brought to bear on our understanding of, and toward the support of, confidence in practical deliberation and judgment. This aim will be pursued through an inquiry into Gadamer's works which focuses on whether we can conceive of a hermeneutically sensitive practical philosophy, that is, a 'hermeneutical philosophy of practice.' Gadamer nowhere uses such a term nor does he explicitly indicate that this is his goal. But as we will show, much of his writing in this field can be put under this heading: he implies that a hermeneutical philosophy of practice is conceivable and that he is engaged in presenting and even defending such a philosophy. Our goal is to advance this project: to make manifest what is latent in Gadamer's

---

5 Hans-Georg Gadamer, "Hermeneutics and Social Science," *Cultural Hermeneutics* 55 (1970): 314, 316. Punctuation in the next to last sentence has been altered to clarify the phrasing.

writings, to clarify its merits in light of the criticisms it has received, to speak through his philosophy where he is reluctant or silent, and to develop the content and application of this practical philosophy beyond the point to which Gadamer has taken it.

To these ends we will take steps to correlate the needs of practical deliberation with the resources of Gadamer's hermeneutical philosophy. We will do so by means of a framework composed of two concepts—*historicity and good*—which is offered as a new and effective way to identify the problem which lies at the root of the contemporary crisis. The meaning of, and relationship between historicity and good are matters that both inform, and are informed by, each of the contrasting paired concepts that arise in this inquiry: universal and particular, truth and method, theory and practice, fact and value. But insofar as these concepts touch on practical judgment, it is the relationship between historicity and good which embraces them all. This relationship represents the master framework of human practice, the dimension without which practice degenerates into something less than genuine human choice.

Today, however, there seem to be few resources to sustain confidence in the compatibility of historicity and good, and many reasons to dismantle it. What we lack are the intellectual tools to restore the credibility of this relationship and to make confidence in it viable again. And as long as we lack such tools the modern deliberative mind will remain painfully unintegrated. What we seek, then, is a new philosophical perspective on historicity and good.[6]

Philosophical hermeneutics can address these needs. Gadamer believes that truth and historicity 'belong' together, and through his investigations intends to show this 'belonging.' The aim in the present work is to argue that such a 'belonging' also applies to good and historicity, and that the account of this relationship imbedded in philosophical hermeneutics can address the contemporary need for a better philosophical perspective on practice.

---

[6] In pursuing this thesis, the aim is not to confuse or equate historicity and the good; this would be a view as philosophically incorrect (and as politically dangerous to good practice) as that implied by the incompatibility thesis. To understand how the compatibility of historicity and good is productive for practical philosophy, we must maintain a sincere sensitivity to the differences between historicity and good as well. The results of historical research which show the contingency of all our claims, as disconcerting as they may be to ethical confidence, are not to be evaded. Nor should we underestimate the manifestations of good, whose mysterious ubiquity deserves respect.

Furthermore, and in some respects beyond Gadamer's own understanding, philosophical hermeneutics requires that we address this need, and that we use the resources of concrete traditions to do so. As this implies, a critique of Gadamer' limitations must be made. But for the first time it is not an extrinsic one, of the kind already made by various critics and based on assumptions drawn from outside his philosophy. The critique developed in the final chapters is an intrinsic one, as are the elements of a solution. Gadamer seems not to have perceived that the insights of his own philosophy address the concerns he evidently harbors about the consequences of actually articulating the kind of practical philosophy envisioned from the hermeneutical perspective. With this barrier lifted, we will have established the future promise as well as present relevance of a hermeneutical philosophy of practice.

ONE

Human Understanding: Interpreting in Place and Time

In its most literal and basic sense, hermeneutics means a set of principles, rules or guidelines—a kind of manual, perhaps—for interpreting a text. More generally, hermeneutics can be regarded as the theoretical study of textual interpretation and of the ways to ensure good interpretation.[1] In the past century, hermeneutics has also gradually became a

---

[1] We can further characterize the kind of event which involves or requires hermeneutical attention in terms of three human activities with which it has historically been associated. First, to say or express something, to proclaim or perform something, whether in words or some other media, which is itself an interpretation of something else. Here, interpretation means the disclosure of something, either in repetition, as in the proclamation of religious scripture or musical performance, or as an original work, which, even so, the creator often regards as something he or she received.

Second, to explain something, which usually means to use language to indicate the meaning of something else; this presupposes that something needs explanation, and not surprisingly, this thing may often be something which was expressed, in the first sense of a hermeneutical event.

Third, to translate something, again usually in language, but not merely *of* language; translation from one language to another concerns ideas, values, customs and perspectives from places or times whose distance from our own is almost a measure in itself of the difficulty that may be involved in effective translation.

Richard Palmer provides the initial characterization of the threefold distinction developed above; he goes on to make this general observation on hermeneutical events: "In all three cases, something foreign, strange, separated in time, space or experience is made familiar, present, comprehensible; something requiring representation, explanation, or translation is somehow 'brought to understanding'—is

subject of philosophical inquiry as greater attention was given to the nature of human understanding itself and to its interpretive character.[2] The milestones of modern Western philosophical inquiry into hermeneutics—the work of Schleiermacher, Dilthey, Husserl, and Heidegger in particular—have marked a steady enlargement and even radicalization of our 'understanding of understanding'.[3]

Closely related to this transformation of the function and enlargement of the scope of hermeneutics was the genesis of what is now called 'historical consciousness,' which has had profound consequences for the modern mind. This consciousness may be described as an awareness that we are in some way unconnected with the past, not simply by distance, but in a radical way broken away from it, and no longer in unconscious continuity with it. And yet, precisely because of this break, we are able to see in a new way the contingency of our present on what went before— which is the more common, and more 'scientific' meaning of the term. It is significant that the first seeds of this consciousness can be found in the Reformation, which was an effort to re-establish a bond with the past. But in the ensuing rise of modern science, this historical consciousness became something different from its original purpose: the more 'objective' the past and its texts are made, the more alienated we are from them.

But if something was lost in this transition from a religiously motivated historical hermeneutic to a scientific concern with the objectivity of history, the transition also provided a positive opportunity. The universality which science claimed in making its study of history and human behavior itself became the object of philosophical scrutiny. The question of universality made available a platform for recovering, in a different way and time, the question of our belongingness to the past which the Reformation had first felt compelled to raise. The issue of hermeneutics could now again become one of meaning in history, not in terms of a single religious tradition alone (which is the task of theological hermeneutics) but in terms of the general human condition of living within cultures and their histories.

'interpreted.'" (Richard Palmer, *Hermeneutics: Interpretation Theory in Schleiermacher, Dilthey, Heidegger, and Gadamer* [Evanston, IL: Northwestern University Press, 1969], 14. See pp. 14-31, on which the preceding discussion is loosely based.)

[2] See Palmer, 35-40.

[3] See Paul Ricoeur, "The Task of Hermeneutics," in *Paul Ricoeur: Hermeneutics and the Human Sciences*, ed. and trans. John B. Thompson (Cambridge: Cambridge University Press; Paris: Editions de la Maison des Sciences de l'Homme, 1981), 43-44.

Following on the development of textual hermeneutics and more recently philosophical hermeneutics, we now seem to be in the opening phases of a third stage which continues this inquiry in the context of the relevance of hermeneutics, and the philosophy of it, to diverse fields in the humanities and social sciences.[4] Gadamer's hermeneutical philosophy can be located between the second and third phases; indeed, the transition to the most recent phase has become possible in no small measure because of his labors. His work is rooted in textual study of the classics, and has matured in the development of his own philosophical perspective which he been in the process of applying to a variety of subject areas, moral practice and moral reason being among them.[5]

---

[4] So far, this includes linguistics, history, aesthetics, Marxian critical theory, philosophy of science, political theory and theology. In fact, hermeneutics is an idea which now appears more and more frequently, even in popular works, and is taking on a life of its own which may or may not be aware of its origins and larger historical character. We can hear of a 'hermeneutics of' this or that, and it is sometimes used loosely and even faddishly, perhaps to give an aura of respectability to what is simply an author's view. But on the whole, the widening recognition of hermeneutics has caused new self-reflection on the nature of our human understanding and has led to exciting inquiries and debates in various circles.

[5] Before proceeding further, it may be helpful to make a few comments on terms:

(1) Gadamer evidently prefers the term "philosophical hermeneutics" to describe the object of his work, although we will use it interchangeably with "hermeneutical philosophy." If there is a difference between the referents of these terms, then both appear in Gadamer's work: on the one hand, he presents a hermeneutic—an interpretation—of human understanding; on the other hand, he thereby creates a philosophy, not a systematic one but a cohesive, reasoned perspective on this subject. In short, it is a philosophy about the hermeneutical event, and a philosophy which itself proceeds hermeneutically.

(2) There may be some confusion about the difference between the two nouns, "hermeneutic" and "hermeneutics"—or lack of difference, especially in view of how interchangable they seem to be in many works. In our view, the plural form refers to a set of principles designed to explain and guide the process of interpretation; the singular form refers to a single interpretation of something, to some result of applying the chosen principles of interpretation. In point of fact, "hermeneutics" is sometimes mistakenly applied to a particular interpretation of a concrete source or text; it would be more accurate here to use the singular form (or even to use "hermeneusis," the Greek noun form of the verb "hermeneuein," to interpret). However, in Gadamer's case—and in other cases as well—one often blurs into the other: the articulation of a principle of interpretation may also contain or imply an interpretation about a particular source. The importance of the difference should be noted, although no sharp distinction is defined or assumed in the present work.

(3) The novice may puzzle over something which the 'expert' has ceased to notice: that Gadamer (and all our sources, for that matter) use singular pronouns in conjunction with the word "hermeneutics"—eg., "This is a hermeneutics of moral-

Perhaps his most basic aim in these efforts is to restore the hermeneutical event to awareness, yet within and not merely against the modern age. Such a restoration does mean, however, opposing the scientific thesis that objectivity is the sole criteria of truth, and exposing as false the more diffuse but pervasive idea of our age, that hermeneutical events can usually be dismissed as marginal and occasional events in the larger arena of human understanding. In fact, Gadamer asserts, confused and partial understanding are more the rule— not one to shame us, but one which has much to tell us about the possibility of discovering ever-new understanding.

There are of course, many aims in Gadamer's works, sometimes seeking expression simultaneously. This can be inferred, for example, from the following passage, which has the merit of being a particularly succinct statement about what he considers the hermeneutical experience to be. It is also significant for his closing referent to relativism, an issue which becomes more central for our inquiry later, and which reminds us that he hopes to discipline as well as contribute to the present, third phase in the expanding and deepening development of attention to the hermeneutical event.

> As I see it, the hermeneutical experience is the experience of the difficulty that we encounter when we try to follow a book, a play, or a work of art step by step, in such way as to allow it to obsess us and lead us beyond our own horizon. It is by no means certain that we can ever recapture and integrate the original experiences encapsulated in those works. Still, taking them seriously involves a challenge to our thinking and preserves us from the danger of agnosticism or relativism.[6]

THE EXPERIENCE OF UNDERSTANDING

Our aim in this chapter is to explore several of these ideas by surveying several primary themes in *Truth and Method*. Here Gadamer lays out his mature perspective on what interpretation is and what it can tell us about human existence itself. This provides a perspective and lexicon on which our inquiry can build.

---

ity"—even though the term itself is plural. The plurality of the term refers to the presumed multiplicity of principles which collectively make up one approach to this activity; by convention such a collection receives singular pronouns.

   [6] Hans-Georg Gadamer, an interview conducted and edited by Ernest L. Fortin, "Gadamer on Strauss: An Interview," *Interpretation* 12, no. 1 (January 1984): 6-7.

In the Introduction to his magnum opus, Gadamer tells us that he seeks to investigate nothing less than the "phenomenon of understanding and . . . the correct interpretation of what has been understood."[7] But an equally important motivation of his work is reconstructive critique. For Gadamer, hermeneutics is antecedent to modern science both philosophically and historically: "From its historical beginnings, the problem of hermeneutics goes beyond the limits of the concept of method as set by modern science."[8] To establish this anew for the modern mind, the effort therefore must be both corrective and constructive.

Gadamer critically argues that "the hermeneutic phenomenon is basically not a problem of method at all. It is not concerned with a method of understanding by means of which texts are subjected to scientific investigation like all other objects of experience."[9] In constructive terms, Gadamer's philosophy of hermeneutics "is concerned to seek the experience of truth that transcends the domain of scientific method wherever that experience is to be found, and to inquire into its legitimacy."[10] These aims are joined together in Gadamer's philosophy as "an attempt to understand what the human sciences truly are, beyond their methodological self-consciousness, and what connects them with the totality of our experience of world."[11]

Hermeneutic phenomena thus are the original subject matter of human science, and Gadamer aims in *Truth and Method* to explore that hermeneutic character in "the experiences of philosophy, of art, and of history itself. These are all modes of experience in which a truth is communicated that cannot be verified by the methodological means proper to science."[12] By these means Gadamer ambitiously seeks "to develop . . . a conception of knowledge and of truth that corresponds to the whole of our hermeneutic experience." Significantly, that experience "not only needs to be justified philosophically, but . . . is itself a way of doing philosophy."[13] In this way his philosophy of hermeneutics is informed by its object, the event of understanding. Hence, he intends to allow his claims to shape and guide the presentation of his philosophy.

---

7   *TM*, xxi [xi].
8   *TM*, xxi [xi].
9   *TM*, xxi [xi].
10   *TM*, xxii [xii].
11   *TM*, xxiii [xiii].
12   *TM*, xxii [xii].
13   *TM*, xxiii [xiii].

Although Gadamer's treatment of the nature of experience has, as he himself puts it, a "key position" in his major work,[14] it appears not in the beginning but the middle of *Truth and Method*, in his section on the "Foundation of a Theory of Hermeneutical Experience." This is the part of his work on which we will focus in this chapter. In this discussion the issue concerns the sorts of understanding which can be associated with or derived from experience. Gadamer argues that experience has a "fundamental openness to new experience" which cannot be accounted for or incorporated by the methodological ideal of repeatability in science, and that this openness must be associated with the "negativity of experience [which] has a curiously productive meaning." If this is unexpected, it may be because, as Gadamer says, the concept of experience today is "obscure," having been subordinated to the requirements of modern science. Repeatability, as the criterion of validity for both the natural sciences and "historico-critical" human sciences, requires "science so to objectify experience that it no longer contains any historical element." This attention to the repeatability of knowledge at the expense of the historical character of experience is a "one-sidedness," which cannot be reformed because this "theory of experience is related exclusively teleologically to the truth that is derived from it."[15] It can only be corrected by starting anew with the nature of experience and the knowledge we gain from it.

Gadamer says that every new experience of an object provides

> better knowledge through it, not only of itself, but of what we thought we knew before—i.e., of a universal. The negation by means of which it achieves this is a determinate negation. We call this kind of experience dialectical.[16]

So experience is productive by its own negative instances, and it is this kind of dialectic, rather than the logic of repeatability, which is the basis of human self-understanding. This dialectic produces knowledge, but what it produces remains subject to the dialectic.

What kind of knowledge does this dialectic produce? Gadamer illustrates the productivity of dialectical experience in the common notion of the experienced person, a notion which exposes the heart of the extra-scientific knowledge he seeks to justify. The experienced person does not

---

14   *TM*, xxxv [xxiii].
15   *TM*, 346-353 [310-317] passim.
16   *TM*, 353 [317].

indicate someone who knows more facts than most people, or who knows 'The Truth.'

> Rather, the experienced person proves to be, on the contrary, someone who is radically undogmatic; who, because of the many experiences he has had and the knowledge he has drawn from them, is particularly well equipped to have new experiences and to learn from them. The dialectic of experience has its proper fulfillment not in definitive knowledge but in the openness to experience that is made possible by experience itself.[17]

Dialectical experience preserves the historical dimension of experience by embracing its negative instances. "Every experience worthy of the name thwarts an expectation. Thus the historical nature of man essentially implies a fundamental negativity that emerges in the relation between experience and insight." Insight—a key concept for Gadamer—is not conventional attributive knowledge of "this or that particular thing," but is the knowledge which extra-scientific experience gives. It is "insight into the limitations of humanity, into the absoluteness of the barrier that separates man from the divine. It is ultimately a religious insight—the kind of insight that gave birth to Greek tragedy."[18] Gadamer's conclusion is evident throughout his writing:

> Thus experience is experience of human finitude . . . The experienced man knows that all foresight is limited and all plans uncertain. In him is realized the truth-value of experience . . . In . . . [experience,] all dogmatism, which proceeds from the soaring desires of the human heart, reaches an absolute barrier. Experience teaches us to acknowledge the real . . .
> . . . To acknowledge what is does not just mean to recognize what is at this moment, but to have insight into the limited degree to which the future is still open to expectation and planning or, even more fundamentally, to have the insight that all the expectation and planning of finite beings is finite and limited. Genuine experience is experience of one's own historicity.[19]

Gadamer proceeds to develop his theory of experience in terms of the structures by which we gain understanding from experience. The primary structure is that of questioning, which is already implicit in experience, where events contradict expectations. More explicitly, "discourse that is intended to reveal something requires that that thing be broken open by the question."[20]

---

17 *TM*, 355 [319].
18 *TM*, 355, 357 [319, 320].
19 *TM*, 357 [320, 321].
20 *TM*, 362, 363 [325, 326].

This openness, however, "is not boundless. It is limited by the horizon of the question." This "implies the explicit establishing of presuppositions, in terms of which can be seen what still remains open." Without this, the dialectic of the question is only one of "fluid indeterminacy." Put another way, the question has its own sense. "The sense of the question is the only direction from which the answer can be given if it is to make sense. A question places what is questioned in a particular perspective. When a question arises, it breaks open the being of the object, as it were."[21] The limitation of the sense of a question does not stifle its openness, but, significantly, is the basis of that openness.

Inattention to this productive limitation results in the failure to pose a true question, one which is open. For Gadamer, the failure to pose an open question can result in either false or slanted questions. "We say that a question has been put wrongly when it does not reach the state of openness but precludes reading it by retaining false presuppositions. It pretends to an openness and susceptibility to decision that it does not have." A slanted question has a different character; it has "a question behind it—i.e., there is an openness intended, but it does not lie in the direction in which the slanted question is pointing . . . It does not give any real direction, and hence no answer to it is possible."[22] Thus, the capacity to pose open questions is the foundation of excellence in knowledge about the human world.

Although our understanding is certainly finite, it is also always larger than the judgment or decision we make about the alternative possibilities. The experience of finitude gives us knowledge that knowledge will always be infinite, always larger than that part of it which we embrace in decision. And this awareness can help guide our questions toward genuine openness. "Hence the logos [of the question] that explicates this opened-up being [of the object] is an answer."[23]

This is, in fact, part of Gadamer's answer to the question of how we can go about asking open questions. There is no method or criterion by which we could guarantee the results of our efforts. Questioning must be understood as an art, as it was in the classical tradition. This art does not guarantee true knowledge or even win all disputes; nor does it avoid "the pressure of opinion; it already presupposes this freedom." It cannot even be taught or learned. Gadamer goes on:

---

21 *TM*, 362, 363 [326, 327].
22 *TM*, 364 [327].
23 *TM*, 362 [326].

As the art of asking questions, dialectic proves its value because only the person who knows how to ask questions is able to persist in his questioning, which involves being able to preserve his orientation toward openness. The art of questioning is the art of questioning ever further—i.e., the art of thinking. It is called dialectic because it is the art of conducting a real dialogue.[24]

In principle, then, there is no barrier between our horizon and new knowledge; the only impenetrable barriers are those which we erect in false questions or in questions that are slanted, in which the intent is distorted. Our presuppositions contain knowledge, and the new knowledge gained by questioning is available as new presuppositions. Complete or finished knowledge, however, is not available. This dialectic is not progressive, in the sense of accumulating knowledge, but processive, in that continually new knowledge is discovered, yet the unknown is never exhausted.

HISTORICAL CONSCIOUSNESS AND THE HUMAN SCIENCES

The theory of experience just discussed forms the foundation of Gadamer's position as he addresses the problem of historical interpretation, of understanding texts from earlier times or different cultures. The problem, in essence, is that while we often want to understand a text in order to understand the time and place from which it came, yet it always seems that we must already know these in order to understand the text. For us today there is also an additional, but related problem: what is the status of our consciousness of history? What does it mean that our awareness that every historical datum can be understood as the product of local, particular forces, and need not—in fact, cannot—justify any absolute or universal inferences?

These matters are of direct relevance to ethics, of course: a principal problem of modern ethics is whether the moral principles we put forward are universally real and true or are relative to the particular contexts in which they appear. Perhaps it is a sign of the times that there seems to be no choice except either absolutism or relativism, or as might be said in the present context of historical rather than moral interpretation, either objectivism or relativism.

---

[24] *TM,* 367 [330].

Gadamer's concern with the definition of scientific self-understanding is structured by an interface between the nature of the *Geisteswissenschaften* (which is usually translated as 'the human sciences' as opposed to the natural sciences)[25] and the meaning of historical consciousness. The issue is a struggle over what will be the fate of the human sciences in a world so very much shaped by this modern phenomenon of historical consciousness. Will our understanding of human science be dominated by objectivism, which, carried over from the natural sciences, characterizes much of the social sciences today? This obsession with objective knowledge excludes another, indispensable kind of knowledge about history and about ourselves. Self-knowledge cannot be 'objective' in the way that natural science measures its experiments, but this does not mean that it does not give us knowledge—it gives reliable knowledge, not 'certain' knowledge.

The history of textual hermeneutics becomes a resource and paradigm by which Gadamer develops his hermeneutic concept of the human sciences and criticizes the limitations and illusions—the false or slanted questioning—which characterize human sciences that have been made over by objectivism. But in developing this philosophical position, he finds an opportunity to reconceive historical consciousness—to pose a new question about it—instead of either simply rejecting it, or viewing relativism as the inevitable but tolerable cost of embracing historical consciousness. There is a rich tradition of philosophical effort, from Hegel to Heidegger, to understand human understanding in terms of its essential historicity, of its 'historicality.' From this perspective, everything about us has location, antecedents and effects, both temporally and culturally, even before we try to determine the epistemological scope or moral meaning of this fact. Gadamer's work represents an effort to rescue historicity from the mistaken prejudices of methodological objectivism and

---

25 *Geisteswissenschaften* literally means sciences of the spirit or mind; Gadamer traces it to the German translator who used it to render J. S. Mill's 'moral sciences' (*TM*, 3 [5]). Elsewhere, Gadamer points out that Mill's intention was not to give the 'moral sciences' their own logic, which characterizes the contemporary intent when we speak of 'the human sciences,' but "to show that the inductive method found at the base of all empirical science is also the only valid method for the domain of the moral sciences" (Hans-Georg Gadamer, "The Problem of Historical Consciousness," *Graduate Faculty Philosophy Journal* 55 [1975]: 12.) Ironically, the English term 'human sciences' is now used to refer to both the humanities and the social sciences, in that both are concerned with human knowledge about what it means to be human, and therefore are both hermeneutic sciences.

historicism, an effort which inevitably will have consequences for our conception of moral and social life.

How, then, does Gadamer speak of historical consciousness and understand its problem? "Having an historical sense" of this kind, he says, "is to conquer in a consistent manner the natural naivete which makes us judge the past by the so-called obvious scales of our current life, in the perspective of our institutions, and from our acquired values and truths."[26] Historical consciousness represents the "privilege of modern man to have a full awareness of the historicity of everything present and the relativity of all opinions."[27] Gadamer strongly emphasizes the impact of this phenomenon on our ethos, and further hints at its ambiguous value:

> The appearance of historical self-consciousness is very likely the most important revolution among those we have undergone since the beginning of the modern epoch. Its spiritual magnitude probably surpasses what we recognize in the applications of natural science, applications which have so visibly transformed the surface of our planet. The historical consciousness which characterizes contemporary man is a privilege, perhaps even a burden, the likes of which has never been imposed on any previous generation.[28]

The methodology of the modern human sciences displays this consciousness in its response toward traditional truths, a response which is called 'interpretation' because so much of what was traditionally accepted as true has today become foreign or even suspect as false. "We always intend by this that the meaning of what is given over for our interpretation is not revealed without mediation, and that we must look beyond the immediate sense in order to discover the 'true' hidden meaning." The ubiquity of this word 'interpretation' today testifies to

---

26 "The Problem of Historical Consciousness," *Graduate Faculty Philosophy Journal* 55 (1975): 8-9. Gadamer gave a series of lectures under this title in France in 1957 which closely parallel a number of important topics in *Truth and Method*. At certain points, these lectures amplify or condense points made in his larger work, and a corresponding use of them is made here. The lectures were first published in French in 1963 as *Le Probleme de la Conscience Historique*; they were translated into English by Jeff L. Close and published with a new (1975) introduction by Gadamer in the *Journal* [55: 1-52]; most recently, the English translation has been republished in *Interpretive Social Science: A Reader*, ed. P. Rabinow and W. Sullivan (Berkeley: University of California Press, 1979), 103-160.

27 "Problem of Historical Consciousness," 8.

28 "Problem of Historical Consciousness," 8.

"the attitude of an entire epoch."[29] This concept of interpretation did not come about easily and its meaning is still caught up in the question of its relationship to the older Cartesian model of epistemology, on which modern science is based.

In *Truth and Method* Gadamer considers at some length the efforts of several thinkers of the nineteenth century, in particular Wilhelm Dilthey, who sought to move beyond the Cartesian framework toward a general hermeneutic perspective. Dilthey hoped to ground objective historical knowledge through a hermeneutic adaptation of *Lebensphilosophie*, a perspective which Gadamer finds in Dilthey and other precursors and labels 'romantic hermeneutics.' Gadamer is interested in Dilthey because he represented an important transitional step in the story of the passage from historicism to a new hermeneutical perspective. But he is also significanct, in Gadamer's view, because he failed to leave behind the Cartesian legacy which will always prevent this passage. It stands as a warning not only to Gadamer's efforts, but to any alternative development or application of hermeneutics which Gadamer's interlocutors might wish to offer.[30]

Gadamer describes his goal as "the task to accomplish the transition from a hermeneutic methodology in the sense of Schleiermacher and Dilthey, to a hermeneutic philosophy in the context of the special position of the *Geisteswissenschaften* relative to the natural sciences."[31] As we shall see, especially in Chapter Three, Gadamer seeks to distinguish his hermeneutical concept of the contingency of historical existence from the historicist concept of this reality.

It was the work of Heidegger who provided the new basis for conceiving human understanding on which Gadamer built his own contribution. Gadamer testifies to Heidegger's importance when he says that, as a result of the 1927 publication of Heidegger's *Being and Time*, "*the whole idea of grounding itself underwent a total reversal*," which "burst asunder the whole subjectivism of modern philosophy."[32] The problem, as Heidegger puts it, is that "any interpretation which is to contribute understanding, must already have understood what is to be interpreted," yet "in a scientific proof, we may not presuppose what it is our task to provide grounds for." How should this problem be addressed? By seeking one or another way of supposedly grounding some independent

---

29   "Problem of Historical Consciousness," 9.

30   "Problem of Historical Consciousness," 23.

31   "Problem of Historical Consciousness," 3.

32   *TM*, 257 [227, 228]. His italics.

knowledge? But the circularity of understanding destroys every such effort. Heidegger's reply, however, completely inverts the supposition of the question. This hermeneutic circle—in which every whole that is offered as a new ground is discovered to have already depended on some part which cannot be grounded—is not a problem from which we must devise an escape:

> But if we see this circle as a vicious one and look for ways of avoiding it, even if we just 'sense' it as an inevitable imperfection, then the act of understanding has been misunderstood from the ground up. The assimilation of understanding and interpretation to a definite ideal of knowledge is not the issue here. Such an ideal is itself only a subspecies of understanding . . . What is decisive is not to get out of the circle but to come into it in the right way . . . It is not to be reduced to the level of a vicious circle, or even of a circle which is merely tolerated. In the circle is hidden a positive possibility of the most primordial kind of knowing.[33]

Gadamer's work presupposes this revaluation of the hermeneutic circle and proceeds from it. In the following passage Gadamer interprets the significance of Heidegger for his own work:

> Against the background of this kind of existential analysis of Dasein [There-being] . . . the problems of a hermeneutics of the human sciences suddenly look very different. The present work is devoted to this new aspect of the hermeneutical problem. In reviving the question of being and thus moving beyond all previous metaphysics—and not just its climax in the Cartesianism of modern science and transcendental philosophy—Heidegger attained a fundamentally new position with regard to the aporias of historicism. The concept of understanding is no longer a methodological concept . . . [34]

This perspective is reflected in the theory of experience outlined above, and it informs the manner in which he develops his theory of textual interpretation, to which we now turn.

HERMENEUTICS AND HISTORY

Gadamer often points out similarities between various forms of communication in order to demonstrate the universality of hermeneutic

---

[33] Martin Heidegger, *Being and Time*, trans. John Macquarrie and Edward Robinson (New York: Harper & Row, 1962), 194. Italics his.

[34] *TM*, 259 [230].

principles. The path connecting conversation, or dialogue, to historical understanding is especially important. By showing their common characteristics, he indicates how the process of interpreting texts has its place within his theory of human experience. In short, he claims that historical understanding is a specialized derivative of the process of conversation.

A dialogue "necessarily has the structure of question and answer," which is the original form of the experience of understanding. One does not so much direct the conversation at a partner as "allow oneself to be conducted by the subject matter to which the partners in the dialogue are oriented." [35]

> Reaching an understanding in conversation presupposes that both partners are ready for it and are trying to recognize the full value of what is alien and opposed to them. If this happens mutually, and each of the partners, while simultaneously holding on to his own arguments, weighs the counter-arguments, it is finally possible to achieve—in an imperceptible but not arbitrary reciprocal translation of the other's position (we call this an exchange of views)—a common diction and a common dictum. [36]

Against the background of this paradigm of understanding in conversation, Gadamer gives the written text special attention because it "has detached itself from the contingency of its origin and its author and made itself free for new relationships." [37] But understanding written texts still reflects the process found in conversation:

> In dialogue spoken language—in the process of question and answer, giving and taking, talking at cross purposes and seeing each other's point— performs the communication of meaning that, with respect to the written tradition, is the task of hermeneutics. Hence, it is more than a metaphor; it is a memory of what originally was the case, to describe the task of hermeneutics as entering into dialogue with the text. [38]

Thus, the task of hermeneutics proper, to guide the interpretation of texts, is one expression of the concrete and universal phenomenon of conversational understanding. In earlier hermeneutical theory, interpretation was thought to be independent of and prior to understanding, and was considered a technical act. However, like the achievement of a com-

---

[35] *TM*, 367 [330].

[36] *TM*, 387 [348]. In the first edition, the last phrase is clearer though more prosaic: "—a common language and a common statement."

[37] *TM*, 395 [357].

[38] *TM*, 368 [331].

mon language in conversation, interpretation in hermeneutics "is not a
means through which understanding is achieved, but . . . has passed into
the content of what is understood." We should conclude that
"understanding and interpretation are indissolubly bound up with each
other."[39] The scope of hermeneutics is thus enlarged beyond its traditional localized meaning of text interpretation, and the mode or process
of that interpretation has been transformed from one of objective recon-
struction to one of dialogue conceptualized as dialectic. This transformation, and its basis in the universality of conversational experience, are
foundations of Gadamer's claim to present a universal philosophy of
hermeneutics.[40]

The experience of understanding a text can now be conceived in this
way:

> A person who is trying to understand a text is always projecting. He projects
> a meaning for the text as a whole as soon as some initial meaning emerges in
> the text. Again, the initial meaning emerges only because he is reading the
> text with particular expectations in regard to a certain meaning. Working
> out this fore-project, which is constantly revised in terms of what emerges as
> he penetrates into the meaning, is understanding what is there.[41]

This affirms the value of the projected or anticipated meaning (the "fore-
project") we bring to interpretation. But focussing on this alone will
obscure what the text means. The critical question is how we 'handle'
such anticipations as we encounter the text.

> A person trying to understand a text is prepared for it to tell him something
> . . . The important thing is to be aware of one's own bias, so that the text can
> present itself in all its otherness and thus assert its own truth against one's
> own fore-meanings.[42]

Clearly, Gadamer is trying to coordinate two facets of interpretation in
these passages, that of listening to what the text really says and that of
relying on one's resources for projecting and revising that message.
Although some may view these two facets of interpretation as irreconcilable, the point is that in any experience of well-done interpretation, they

---

[39] *TM*, 398, 399 [359, 360].

[40] Gadamer wrote in 1975 that in *TM*, "I have tried to describe . . . how this process
of challenge mediates the new by the old and thus constitutes a communicative process built on the model of dialogue. From this I derive hermeneutic's claim to universality." ("Introduction" to "Problem of Historical Consciousness," 6.)

[41] *TM*, 267 [236].

[42] *TM*, 269 [236].

are both already present. "All that is asked is that we remain open to the meaning of the other person or text. But this openness always includes our situating the other meaning in relation to the whole of our own meanings or ourselves in relation to it."[43] When Gadamer says that we should be aware of our bias and our prejudices, he does not mean we do so in order that we may then ignore them, but that we may travel a path where we project our interpretation, test and revise it. On the one hand, the openness required of the interpreter is not empty, just as the openness of the question in everyday conversation is not empty. On the other hand, this newness, this awareness that the text presents something foreign, is the beginning of understanding because it initiates a dialogue through which we move toward fuller understanding.[44]

The Conditions of Interpretation as its Resources

This description of the experience of interpretation represents a new way of approaching the central practical question of any theory of interpretation, which is how to avoid or minimize misunderstanding. To more fully analyze the hermeneutic event in the context of this aim, Gadamer rehabilitates positive meanings of four concepts which have often had pejorative or negative connotations: prejudice, authority, tradition and temporal distance. Gadamer shows both that the conditions which characterize the circle of interpretation are unavoidable, and that since understanding does occur with and through these conditions, they are in fact productive resources for the hermeneutic task. Not surprisingly, questions and criticisms of these concepts are central in the objections posed by Gadamer's critics, but they also pose problems for sympathetic commentators. In later chapters these concepts will be taken up again.

What was thought to be the occasional problem of misunderstanding now appears as the universal problem of *prejudice*. By casting his discussion in term of the value-charged idea of prejudice, Gadamer means to directly and dramatically challenge the Enlightenment's repudiation of it.

---

43   *TM*, 268 [238].
44   All of this goes into Gadamer's own appropriation of the so-called 'hermeneutical circle': "Thus the movement of understanding is constantly from the whole to the part and back to the whole. Our task is to expand the unity of the understood meaning centrifugally. The harmony of all the details with the whole is the criterion of correct understanding." (*TM*, 291 [259].)

This recognition that all understanding inevitably involves some prejudice gives the hermeneutical problem its real thrust. In light of this insight it appears that *historicism, despite its critique of rationalism and of natural law philosophy, is based on the modern Enlightenment and unwittingly shares its prejudices.* And there is one prejudice of the Enlightenment that defines its essence: the fundamental prejudice of the Enlightenment is the prejudice against prejudice itself, which denies tradition its power.[45]

Gadamer intends to recover a non-pejorative sense of 'prejudice,' in which it does not necessarily mean a false or baseless judgment. Drawing on the history of concepts in jurisprudence, Gadamer shows that "actually 'prejudice' means a judgment that is rendered before all the elements that determine a situation have been finally examined." Hence a prejudiced judgment is a "provisional decision."[46] Since, for Gadamer, in the context of the human sciences, we can never finally and conclusively examine a situation, every judgment is seen as prejudiced because it is provisional. Thus, to speak of prejudice in interpretation is for Gadamer a required acknowledgement of what actually happens. It also invites a redefinition of the hermeneutical problem:

> If we want to do justice to man's finite, historical mode of being, it is necessary to fundamentally rehabilitate the concept of prejudice and acknowledge the fact that there are legitimate prejudices. Thus we can formulate the fundamental epistemological question for a truly historical hermeneutics as follows: what is the ground of the legitimacy of prejudices? What distinguishes legitimate prejudices from the countless others which it is the undeniable task of the critical reason to overcome?[47]

For Gadamer it is *authority* which represents legitimate prejudice.[48] In terms of textual interpretation, this means affirming that we "are fundamentally open to the possibility that the writer of a transmitted text is better informed than we are, with our prior meaning."[49] By arguing that an authority is not necessarily illegitimate because it is based on a prejudice or prejudgment, Gadamer contradicts the Enlightenment's judgment of all prejudices as illegitimate, a judgment which results in a dichotomy between critical reason and traditional authority. The implication of that dichotomy is that reason leads to true knowledge and therefore to histor-

---

45 *TM*, 270 [239-240]. His italics.
46 *TM*, 270 [240].
47 *TM*, 277 [246].
48 See *TM*, 278-281 [247-249].
49 *TM*, 284 [262].

ical freedom, while authority is based on false, untested or undemon-
strable knowledge, and requires the 'blind obedience' which has historically oppressed humanity so often.

Gadamer distinguishes legitimate authority from blind obedience:
authority "rests on acknowledgement and hence on an act of reason itself
which, aware of its own limitations, trusts to the better insight of
others."[50] Gadamer also says this: "True prejudices must still finally be
justified by rational knowledge, even though the task can never be fully
completed."[51] But if a prejudice (a prejudgment) has been rationally
justified, need we or should we still consider it a prejudice? As we will
later see, this question implies one of Habermas' objections to this concept. Here it can be noted that Gadamer's emphasis lies on the final
phrase of the above quotation, which is reiterated at the close of his book
when he says, "there is undoubtedly no understanding that is free of all
prejudices."[52] We might add that those prejudices which we most value,
such as great moral ideals and virtues, are those we can least expect to be
ever justified in the sense of proving them beyond the 'shadow' of
prejudice.

Tradition as a source of authority is certainly a contentious concept
today. Gadamer notes that the concept of tradition as a social authority
whose legitimacy lies beyond reason was introduced by Romanticism as
a rallying point in opposition to the Enlightenment, which in turn
viewed tradition as a storehouse of unfounded and often concealed
prejudices. However, Gadamer regards both such conceptions of tradition as distortions of our real relation to the past. We are not, as we often
assume, in a position of nearly complete independence from the past:

> Rather, we are always situated within traditions, and this is no objectifying
> process—i.e., we do not conceive of what tradition says as something other,
> something alien. It is always part of us, a model or exemplar, a kind of
> cognizance that our later historical judgment would hardly regard as a kind
> of knowledge but as the most ingenuous affinity with tradition.[53]

By no means does Gadamer accept every particular tradition as legitimate. But almost always he speaks of it in a universal sense, which frustrates efforts to find out more specifically and substantively what he
means by it, as we will later pursue. As for the relation between reason

---

50   *TM*, 279 [248].
51   *TM*, 273 [242].
52   *TM*, 490 [446].
53   *TM*, 282 [250].

and tradition, Gadamer perceives them to be in close relation, and explains this in an important passage:

> It seems to me, however, that there is no such unconditional antithesis between tradition and reason . . . The fact is that in tradition there is always an element of freedom and of history itself. Even the most genuine and pure tradition does not persist because of the inertia of what once existed. It needs to be affirmed, embraced, cultivated. It is, essentially, preservation, and it is active in all historical change. But preservation is an act of reason, though an inconspicuous one . . . Preservation is as much a freely chosen action as are revolution and renewal.[54]

The human sciences are not guided by the concept of cumulative knowledge that is found in the natural sciences; instead, it is indicative of the unique character of the human sciences that its fruits "almost never become outdated."[55] The human sciences, far from escaping tradition, in fact depend on it, and therefore we must reconsider the nature and method of human science:

> At the beginning of all historical hermeneutics, then, *the abstract antithesis between tradition and historical research, between history and the knowledge of it, must be discarded.* The effect (Wirkung) of a living tradition and the effect of historical study must constitute a unity of effect, the analysis of which would reveal only a texture of reciprocal effects.[56]

Temporal distance is what separates the interpreter and a text which comes from a different age and culture. Although it would seem to be a barrier to or destroyer of historical understanding (and no doubt in certain ways it is), the hermeneutic circle we share with tradition gives a new, positive significance to temporal distance. It is not "a gulf to be bridged," as historicism asserts, but rather, "it is actually the supportive ground of the course of events in which the present is rooted." This process has "genuine productivity," and therefore is constitutive for all understanding.[57] Temporal distance

> lets the true meaning of the object emerge fully . . . Not only are fresh sources of error constantly excluded, so that all kinds of things are filtered out that obscure the true meaning; but new sources of understanding are

---

54 *TM*, 281-282 [250].

55 *TM*, 284 [252].

56 *TM*, 282-283 [250-251]. His italics, and the translator's parenthetical insertion of *Wirkung*.

57 *TM*, 297 [264, 265].

continually emerging that reveal unsuspected elements of meaning . . . And along with the negative side of the filtering process brought about by temporal distance there is also the positive side, namely the value it has for understanding. It not only lets local and limited prejudices die away, but allows those that bring about genuine understanding to emerge clearly as such.[58]

Implications of the Hermeneutical Event

But how can this be possible when prejudices operate in our thinking already and unconsciously? This is a problem with which the Enlightenment also struggled in its desire to start from a scientifically certain point of knowledge. But Gadamer's answer, recalling his discussion of the experienced person, rejects the goal of nullifying prejudices before investigation begins. Instead he argues that we become conscious of a prejudice when it is stimulated, and that the temporally distant text or foreign meaning can have this effect. This is simply another identification of the structure of questioning, and in yet another way Gadamer's theory of experience is brought to bear on the process of historical hermeneutics.

> Understanding begins . . . when something addresses us . . . We now know what this requires, namely the fundamental suspension of our own prejudices. But all suspension of judgments and hence, a fortiori, of prejudices, has the logical structure of a *question*.[59]

The prejudice therefore is not suppressed, but is presented before our judgment in the form of a question about its own validity or meaningfulness. Again we see the double character of questioning: it arises because prejudices are challenged by the foreign text, while to pose the question means to hold the prejudice in obeyance while we wait for an answer. In another passage which elaborates on this role of prejudice, Gadamer employs the term 'play' in a way which evokes both of its meanings, as the exercise of powers and as entering a game.

> In fact our own prejudice is properly brought into play by being put at risk. Only by being given full play is it able to experience the other's claim to truth and make it possible for him to have full play himself.[60]

---

58 *TM*, 298 [265-266].

59 *TM*, 299 [266].

60 *TM*, 299 [266]. Also see Gadamer's "Afterword" to *TM*, 2nd ed.: Since the hermeneutical experience "includes everything intelligible . . . it forces the interpreter to play with his own prejudices at stake." (568)

This surprising conclusion flies in the face of all that scientific methodology teaches about neutrality and objectivity. Instead of trying to excise or set aside our prejudices, Gadamer tells us that we should permit them to be stimulated through encounter with what is strange or unfamiliar, and thereby risk them—risk their repudiation, modification, maturation or substitution. In this light, not only temporal distance but all kinds of strangeness and limitation become part of the possibility of new understanding: the stimulation of our prejudgments makes them accessible to us and permits us to form a question out of their collision with what is foreign. In this way Gadamer finds Heidegger's concept of the hermeneutic circle vindicated in historical understanding:

> Heidegger describes the circle in such a way that the understanding of the text remains permanently determined by the anticipatory movement of fore-understanding. The circle of whole and part is not dissolved in perfect understanding but, on the contrary, is most fully realized.
>
> The circle, then, is not formal in nature. It is neither subjective nor objective, but describes understanding as the interplay of the movement of tradition and the movement of the interpreter. The anticipation of meaning that governs our understanding of a text is not an act of subjectivity, but proceeds from the commonality that binds us to the tradition.[61]

For his own part, Gadamer draws several further, and important, implications from his analysis of the hermeneutics of historical texts; although he does not systematically identify them, they can be briefly itemized. First, the meaning of a text is not exhausted by the author's intent:

> The real meaning of a text, as it speaks to the interpreter, does not depend on the contingencies of the author and his original audience. It certainly is not identical with them, for it is always co-determined also by the historical situation of the interpreter and hence by the totality of the objective course of history.[62]

Second, the independence of the text from its author does not mean that the interpreter's conclusions about the meaning of a text is better than the author's intended meaning, only that it is different. "It is enough to say that we understand in a *different* way, *if we understand at all.*" He amplifies: "Not just occasionally but always, the meaning of a text goes beyond its author. That is why understanding is not merely a reproduc-

---

61 *TM*, 293 [261].
62 *TM*, 296 [263].

tive, but always a productive activity as well."[63] Third, and consequently, "the discovery of the true meaning of a text . . . is never finished; it is in fact an infinite process."[64] This does not mean we never gain a true understanding of the text, but that its truth is not exhausted by any one interpretation. Fourth, and perhaps most significant for Gadamer's opposition to the prejudice of value-neutrality in the human sciences, understanding is not only a grasping of what the author says, nor a receiving of it "as a mere expression of life." Rather, as Gadamer emphatically asserts more than once, in the act of interpretation the text "is taken seriously in its claim to truth."[65] The truth of this 'thing itself' is in contention.

HERMENEUTICAL EXISTENCE

We began with Gadamer's theory of experience as the knowledge of finitude gained from the dialectic of dialogue, and have examined the conditions, resources and implications of interpreting historical interpretation from that perspective; we move now another step in the direction of his philosophical structure to elaborate briefly on a few specific concepts. Although Gadamer's discussion in *Truth and Method* is not always systematically organized, in the following he expresses the procedure we have followed, as well as introduces the new concept to which we now turn: "The dialectic of question and answer disclosed in the structure of hermeneutical experience now permits us to state more exactly what kind of consciousness historically effected consciousness is."[66]

Consciousness open to the effects of history

The above heading is one of several ways, perhaps the best way, of rendering the German term Gadamer has coined, *wirkungsgeschichtliches*

---

[63] *TM*, 296, 297 [264].

[64] *TM*, 298 [265].

[65] *TM*, 297 [264]. See also 292-293 [259-260], in which this appears: "When we try to understand a text, we do not try to transpose ourselves into the author's mind but . . . we try to transpose ourselves into the perspective within which he has formed his views. But this simply means that we try to understand how what he is saying could be right. If we want to understand, we will try to make his arguments even stronger."

[66] *TM*, 377 [340].

Bewußtsein.[67] It is a short-hand label, in effect, for what Gadamer sees as the content and thrust of a hermeneutically sensitive consciousness. But it has a double meaning, which Gadamer points out in 1965 in his "Forward to the Second Edition" of *Truth and Method*:

> Hence there is a certain legitimate ambiguity in the concept of histori-
> cally effected consciousness . . . as I have employed it. This ambiguity is that
> it is used to mean at once the consciousness effected in the course of history
> and determined by history, and the very consciousness of being thus
> effected and determining.[68]

Thus the concept refers both to a 'consciousness *of* history's effects,' and a 'consciousness effected *by* history.' We will adapt our rendition of the term to suit the meaning being emphasized in each case.

Gadamer leads us to an understanding of this term in a prominent passage where he indicates three ways in which we can approach the 'I-Thou' experience, that paradigm of dialogical understanding which for Gadamer embraces the whole range of interpretive experiences.[69] He marks the last of these three, i.e., historically effected consciousness, as the best type of hermeneutical awareness and practice. This comparison follows his discussion of human experience, in which he concludes that "as a genuine form of experience it [i.e., historically effected experience] must reflect the general structure of experience. Thus we will have to seek out in *hermeneutical experience* those elements that we have found in our analysis of experience in general."[70]

Those three ways of experiencing a 'Thou' in everyday life are explored in terms of their parallels or consequences in the realm of hermeneutical experience. First, we can experience other persons in order to discover regularities of human behavior with the aim of predicting it. In moral terms, Gadamer asserts, this makes persons into means

---

67   According to Joel Weinsheimer and Donald G. Marshall, this rendering comes from Paul Ricoeur ("Translators' Preface," in *TM*, 2nd ed., xv.). For comparison, the 1st English edition of *TM* uses a dense but more literal rendition, "effective-historical consciousness" (e.g., [305]), and in the 2nd revised English edition, Weinsheimer and Marshall follow P. Christopher Smith and replace this with a clearer phrase, "historically effected consciousness" (e.g., 341). This latter rendition, however, connotes only one of the two meanings of the term which Gadamer explicitly identifies, as indicated in the text which follows.

68   *TM*, xxxiv [xxi-xxii].

69   *TM*, 358 [321], where he also says, "For tradition is a genuine partner in dialogue, and we belong to it, as does the I with a Thou."

70   *TM*, 357-358 [321]. His italics.

and destroys their character as ends in themselves. The "equivalent" of this pursuit of generalized human nature in hermeneutical experience is the "naive faith in method and in the objectivity that can be attained through it. Someone who understands tradition in this way makes it an object—i.e., he confronts it in a free and uninvolved way—and by methodically excluding everything subjective, he discovers what it contains." This is the attitude of the modern social sciences, in which the researcher cuts himself or herself off from the very traditions that would make his or her work meaningful, and "flattens out the nature of hermeneutical experience."[71]

In a second form of interpretive experience, "the Thou is acknowledged as a person, but . . . is still a form of self-relatedness . . . This relation is not immediate but reflective." This always permits one to dominate the other: "One claims to know the other's claim from his point of view and even to understand the other better than the other understands himself." While this way is "more adequate" than a knowledge aimed simply at calculating how another will act, it still conceals the "dialectic of reciprocity that governs all I-Thou relationships." As a result, "the claim to understand the other person in advance functions to keep the other person's claim at a distance."[72]

This relation to the 'Thou' "is what we generally call *historical consciousness*." It does not view encounters with the past in terms of general rules but of the unique. Yet, "by claiming to transcend its own conditionedness completely in knowing the other, it is involved in a false dialectical appearance, since it is actually seeking to master the past as it were." This aim of historical consciousness is an illusion, an "unattainable ideal." To imagine that all of one's prejudices have been removed, in fact, results instead in removing oneself from the tradition, the very tradition by which any understanding of the Thou is possible. To act this way in an interpersonal context "changes [the] relationship and destroys its moral bond." And in hermeneutical experience, this perspective *"destroys the true meaning of this tradition in exactly the same way."*[73]

Finally, there is a "third, and highest, type of hermeneutical experience: the openness to tradition characteristic of hermeneutically effected consciousness." In human experience "the important thing is, as we have seen, to experience the Thou truly as a Thou—i.e., not to overlook his

---

[71]   *TM*, 358-359 [322].
[72]   *TM*, 359, 360 [322, 323].
[73]   *TM*, 360 [323, 324]. His italics.

claim but to let him really say something to us." In a characterization with clearly moral undertones, Gadamer says that this openness constitutes the basis of each "genuine human bond." This does not mean we become a slave to the other; but it does include "recognizing that I myself must accept some things that are against me, even though no one else who forces me to do so." Gadamer generalizes from this to our relationship to tradition as a whole: "I must allow tradition's claim to validity . . . in such a way that it has something to say to me."[74] Gadamer contrasts this approach with that of historical consciousness:

> Someone who is open to tradition in this [hermeneutical] way sees that historical consciousness is not really open at all, but rather, when it reads its texts "historically," it has always thoroughly smoothed them out beforehand, so that the criteria of the historian's own knowledge can never be called into question by tradition . . . By contrast, hermeneutically effected consciousness rises above such naive comparisons and assimilations by letting itself experience tradition and by keeping itself open to the truth claim encountered in it. The hermeneutical consciousness culminates, not in methodological sureness of itself, but in the same readiness for experience that distinguishes the experienced man from the man captivated by dogma.[75]

Gadamer's critique of Dilthey and of romantic hermeneutics in general is, in effect, that they fail to conceive of the third kind of hermeneutical experience, and that the reason is their attachment to the objectivist pretensions of the second form, the historical consciousness of today's human sciences. Using Heidegger's work as a foundation, and by clearly linking the interpretation of tradition to the character of basic human experience, Gadamer's central effort is to secure this movement from the second to the third type. We may regard consciousness of historical effects as Gadamer's conception of a corrected and authentic historical consciousness, one which finds that acknowledgement of its own historicality "does not limit the freedom of knowledge but makes it possible."[76] This is contrary to what the Enlightenment teaches us, and Gadamer elaborates further on this contrast:

> The naivete of so called historicism consists in the fact that it does not undertake this [hermeneutically best] reflection, and in trusting to the fact that its procedure is methodical, it forgets its own historicity . . . Real historical thinking must take account of its own historicity. Only then will it cease

---

74   *TM*, 361 [324].
75   *TM*, 361-362 [325].

to chase the phantom of a historical object that is the object of progressive research, and learn to view the object as the counterpart of itself and hence understand both . . . *Understanding is, essentially, a historically effected event.*[77]

What implications does Gadamer draw from this analysis for how we should practice interpretation? Gadamer acknowledges that contemporary historians do consider not only an event, but also its effect on subsequent history as well. But inquiry into those effects has been "generally regarded as a mere supplement to historical inquiry," while Gadamer asserts that we must "require an inquiry into history of effect" so that each component of tradition "can be seen clearly and openly in terms of its own meaning." This "new demand" is not one which historical research must take up as a separate task, but is "of a more theoretical kind," "addressed not to research, but to its methodological consciousness." That is, "historical consciousness must become conscious" that "we are always already affected by history." By contrast, "historical objectivism conceals the fact that historical consciousness is itself situated in the web of historical effects," and thus "falls short of reaching that truth which, despite the finite nature of our understanding, could be reached."[78]

Gadamer calls the understanding gained through hermeneutically effected relationships a 'fusion of horizons.' 'Horizon' reflects the structure of our finite consciousness of historical effects. It refers to the scope of one's situation, of one's limited understanding of that situation. The hermeneutic task is not to transcend all horizons, but to achieve "the right horizon of inquiry for the questions evoked by the encounter with tradition." Our horizon should never be regarded as fixed or closed to movement. Neither can we transpose ourselves into an alien horizon nor discard our own. Rather, "we must always already have a horizon in order to be able to transpose ourselves into a situation" which presents us with a different horizon.[79] "*Understanding is always the fusion of these horizons supposedly existing by themselves.*" The effect of this is the "rising to a higher universality that overcomes not only our own particularity but also that of the other." We can see then that "the horizon of the present is continually in the process of being formed because we are continually having to test all our prejudices."[80] In fact,

---

[76] *TM*, 361 [324].
[77] *TM*, 299-300 [266-267]. His italics.
[78] *TM*, 300, 301 [267, 268] passim.
[79] *TM*, 302, 305 [269, 271].
[80] *TM*, 305-306 [272-273] passim. Italics his.

projecting a historical horizon . . . is only one phase in the process of understanding . . . [For] as the historical horizon is projected, it is simultaneously removed. To bring about this fusion in a regulated way is the task of what we called historically effected consciousness.[81]

The implied tension between the conscious and unconscious effects of history is brought out even more in these important lines:

We are not saying, then, that history of effect must be developed as a new independent discipline ancillary to the human sciences, but that we should learn to understand ourselves better and recognize that in all understanding, whether we are expressly aware of it or not, the efficacy of history is at work.[82]

There is a vital interplay between human consciousness and the effects— the workings—of history: on the one hand, unawareness of historical effects can lead to the kind of "naive faith in scientific method" which results in a "deformation of knowledge"; but on the other hand, "the power of effective history does not depend on its being recognized," for "it prevails even where faith in method leads one to deny one's own historicity."[83]

But there is a second level of this interplay which characterizes the effort to be aware of effective history. On the one hand, "our need to become conscious of effective history is urgent because it is necessary for scientific consciousness" (by which Gadamer presumably means science in its legitimate and largest sense). On the other hand, "this does not mean it can ever be absolutely fulfilled;" in fact, it "can never be completely achieved."

The illumination of this situation—reflection on effective history—can never be completely achieved; yet the fact that it cannot be completed is due not to a deficiency in reflection but to the essence of the historical being that we are. *To be historically means that knowledge of oneself can never be complete.*"[84]

Building on the points regarding textual interpretation itemized earlier, let us summarize Gadamer's more philosophical points here, since they will be of critical importance in future chapters. First, effective history is always operative, whether we are aware of it or not; second, to ignore or deny it can result in distorted knowledge or leave prejudg-

---

81 *TM*, 306-307 [273-274].
82 *TM*, 301 [268].
83 *TM*, 301 [268].
84 *TM*, 301, 302 [268, 269]. His italics.

ments unchallenged; third, one should seek to understand the effects of history in one's self-understanding and understanding of a 'Thou'; fourth, one cannot, however, gain complete understanding of the effects of history.

This leaves the consciousness of historical effects in a tenuous position as far as certainty of our acquisition of knowledge, or the clarity of its practical relevance, is concerned. The limits of our understanding may be due either to a wrong approach, to a prejudice, or to inevitable limits of human knowledgeability itself. But this tenuousness is quite to the point: it opposes our expectation of a theory which will guarantee the results of our interpretive practice. Gadamer gives no definite criteria to identify the causes of misunderstanding, or to know when we have gained as much true understanding as is possible. But these problems, which bear on every particular research inquiry, are not, in his view, the concern of a general theory such as his. He simply says: "to acquire an awareness of a situation is . . . always a task of peculiar difficulty."[85] Gadamer has given us tools, but they are not tools which contain the wisdom of how to apply them. Gadamer does, however, offer one other concept which augments our grasp of the act and scene of interpretation.

Application

In the context of historical interpretation, the concept of application is closely related to both historically effected consciousness and the fusion of horizons. Application is the principal act of this consciousness, and the fusion of horizons is its desired end. But application is also a concept, like prejudice or tradition, that has a specific contemporary meaning which conceals the meaning Gadamer wishes to restore and offer to us. But in this case, the contemporary meaning of application is neutral or even positive in connotation: application is associated with the execution of a judgment or act in conformity with the principles or rules of some theory. Gadamer does not quarrel with the occasions where this meaning is legitimate; his point, rather, is to show that application has another meaning, particularly relevant to the human sciences, and that in this sense it is inherently involved in understanding and not an act subsequent to it. As we will see in the next chapter, this sense of application originates, for Gadamer, in Aristotle's ethics, and therefore has particular relevance to our concern with moral and political practice. But here we

---

[85] *TM*, 301 [268-269].

introduce this concept apart from that context, and focus on Gadamer's discussion of its historical legacy in the field of textual interpretation.

Gadamer describes the task of a consciousness of historical effects as "the problem of *application*, which is to be found in all understanding."[86] In pre-modern hermeneutical theory, application was, along with explanation and interpretation, one of the three elements of an act of understanding. But in the hands of the romantics, an emphasis on "the inner fusion of understanding and interpretation" caused application to become "wholly excluded from any connection with hermeneutics." Gadamer's aim is to rehabilitate the significance and role of application in hermeneutics. We are, he says, "forced to go one step beyond romantic hermeneutics, as it were, by regarding not only understanding and interpretation, but also application as comprising one unified process."[87]

The recovery of application in a theory of hermeneutics requires attention to the legal and theological branches of hermeneutics, from which literary and historical hermeneutics "cut their ties" in the pursuit of a "methodology for research in the human sciences." For it is in these other branches that the meaning of application is retained and which thus guide our reconception of hermeneutics in the human sciences. The interpretation of law and scripture depends upon an "essential tension between the fixed text—the law or the gospel—and . . . the sense arrived at by applying it at the concrete moment of interpretation, either in judgment or in preaching." Gadamer goes so far as to say that "understanding here is always application."[88] He elaborates on this, explaining what application means for the interpretation of an historical text:

> Hermeneutics in the sphere of philology and the historical sciences is not "knowledge as domination" . . . but rather, it consists in subordinating ourselves to the text's claim to dominate our minds. Of this, however, legal and theological hermeneutics are the true model. To interpret the law's will or the promises of God is clearly not a form of domination but of service. They are interpretations—which includes application—in the service of what is considered valid. Our thesis is that historical hermeneutics too has a task of application to perform, because it too serves applicable meaning, in that it explicitly and consciously bridges the temporal distance in time that

---

86 *TM*, 274.
87 *TM*, 308 [274, 275].
88 *TM*, 308, 309 [275, 277].

separates the interpreter from the text and overcomes the alienation of meaning that the text has undergone.[89]

Gadamer says he is "quite aware" that he is "asking something unusual of the self-understanding of modern science." Although an affront to the contemporary self-understanding of the human sciences, this demand for a consciousness of effective history is, in a sense, a corrective to the contemporary situation in which the functions of hermeneutics have been separated from each other.[90] Gadamer sums up his conclusions on this point, which parallel his view of the relationship between tradition and rational research into history, and which have relevance to our inquiry into hermeneutics and practice:

> Our line of thought prevents us from dividing the hermeneutic problem in terms of the subjectivity of the interpreter and the objectivity of the meaning to be understood. This would be starting from a false antithesis that cannot be resolved even by recognizing the dialectic of subjective and objective. To distinguish between a normative function and a cognitive one is to separate what clearly belong together.[91]

SUMMATION AND PROSPECT

We have seen that for Gadamer the elimination of misunderstanding is part of the process of understanding, not an aberration or exception to the "normal" process because of any failure to adequately suppress our prejudices beforehand. The demand that we suppress those prejudices is dictated by modern science. Yet even those in the human sciences who obey that dictation acknowledge that objectivity is always merely approximated and not a proven achievement.

---

[89] *TM*, 311 [278].

[90] *TM*, 309, 310 [276, 277]. A notable case of this is Emilio Betti's theory of hermeneutics, about which Gadamer makes some critical comments. Following the traditional view of the functions of hermeneutics, Betti argues that it has cognitive, normative, and reproductive functions. Gadamer disagrees that these functions are separable, and argues that each hermeneutical discipline in fact uses more than just one function. Neither theological nor legal hermeneutics are only normative, but are cognitive (i.e., explanatory) tasks as well. Conversely, artistic interpretation is not simply cognitive but normative as well (*TM*, 309-310 [276]; Gadamer cites Emilio Betti, *Teoria generale dell' interpretazione*, 2 vols. (Milan: A. Guiffre, 1956).

[91] *TM*, 311 [277].

This state of affairs is explained by Gadamer's philosophy. First, the prejudgments which characterize our horizon and shape our questioning cannot be eliminated but in fact are required if inquiry is to be real. Inquiry reflects the exchange between the familiar and the new; it brings horizons into dialogue and even into fusion. Second, the ideal of objectivity discloses a misplaced openness to the new; the openness it tries to have is aborted because the imagined objective or neutral point has no location or center, and no audience to address. Thus, as a methodological ideal, absolute objectivity is simply not relevant to the human sciences.

By contrast, authentic work in the human sciences is characterized by the persistence of the dialectic of questioning the continual fusion of horizons which Gadamer describes. In this process there is no a priori principle to distinguish reason and prejudice. There is no clear dichotomy between true science and legitimate tradition, or between explanation and interpretation, or between interpretation and application. Such distinctions can, however, serve a purpose, though it is secondary or derivative: to be relative controls in a process which ultimately is not controllable. This is the historical character of human self-knowledge.

But this historicity is necessarily more than a condition of knowledge, more than an epistemological principle. Gadamer's claim is that everything understood is by nature dialectical, as well as arrived at dialectically. This is the only possible meaning of Gadamer's statements that "genuine experience is experience of one's own historicity," and that "knowledge is dialectical from the ground up."[92] Gadamer's claim for the hermeneutical character of knowledge is ontological as well as epistemological. If the mode of human being is essentially one of understanding, it can only be the understanding which we have as historical creatures.

But historical creatures are also creatures of practice, in the classical sense of moral and political action. In the chapters that follow we will at first indirectly explore, and then more explicitly address, what implications this ontological dimension of Gadamer's philosophy of hermeneutics has for human practice. For this reason we here depart from an explicit concern with textual hermeneutics except as it relates to our understanding of practice. However, as the acts of writing and reading are themselves forms of practice, our new course will never be far from the genesis of hermeneutics in the interpretation of texts.

---

[92] *TM*, 357, 365 [321, 328].

TWO

Hermeneutics and Practical Philosophy

What is the significance of Gadamer's philosophy of hermeneutics for the activity of understanding? Gadamer certainly presents us with an 'understanding of understanding,' a perspective that starts before and takes us beyond the task of textual interpretation. The root of understanding, Gadamer tells us, is the experience of dialogue and conversation.

But what does this tell us about how we ought to seek understanding? As we saw, Gadamer is definitely recommending cultivation of a 'hermeneutical consciousness,' an awareness of the effects of history on our understanding of the world. Yet no criteria for this consciousness are indicated, and consequently no criteria of truth or true understanding are indicated. Is there nothing more substantive or normative to Gadamer's recommendation than, as David Hoy puts it, an affirmation of an "openness to more experience and the recognition of finitude"? Valuable as it is for "exposing the self-deception of dogmatism,"[1] the problem, which various critics have raised, is

> whether this openness can serve as a criterion for judging the *results* of understanding and interpretation. If openness is to be distinguishable from mere weakness of will, there must be some explanation not only of the tolerance of understanding, but also of its truth. Although understanding is

---

[1] David Hoy, *The Critical Circle: Literature, History and Philosophical Hermeneutics* (Berkeley: University of California Press, 1978), 60-61.

always grounded in a situation, it attempts to rise above that situation in
order to say something true about it.[2]

Without criteria for true understanding, critics "see in Gadamer's
hermeneutics the danger of historical relativism."[3] If this is the case,
hermeneutics will not be able to help us move from an abstract notion of
tradition as the universal human heritage to a judgment of the value of
diverse, concrete traditions in culture or science. It would seem, in this
case, to be unable to make its universals (and thus its claim to universality) applicable to concrete particulars. In short, if philosophical
hermeneutics is posed as lacking norms of judgment or criteria of truth,
does it not end up as isolated from historical life as the methodologism
whose claim to universality Gadamer so emphatically rejects?

These questions will continue to arise in our investigation of
Gadamer's attention to moral practice and moral theory. Following
Aristotle, Gadamer views these areas as providing fundamental
guidance for the activity of all the human sciences. But Gadamer's
response to questions about the criteria of judgment, whether in
interpretive understanding or moral action, do not conform to the
presuppositions or expectations with which his critics pose their
questions, and, indeed, in which the modern public tends to consider
these issues. We will see more than once that Gadamer seems to both
encourage and cancel expectations, to both offer and withhold answers.
Nevertheless, we will be able to discern the compass bearing which
guides his apparently wandering course.

This chapter has three sections. The first section raises the problem of
whether philosophical hermeneutics can have normative relevance to the
human sciences in general, and sets forth Gadamer's unique claim that in
our age this philosophy can serve in some regard the function that
Aristotle once envisioned for practical philosophy and so demonstrate its
relevance. The second section explicates the origins of this claim in
Gadamer's discussions in *Truth and Method* of the affinity between interpretive understanding and moral reasoning. The third section further
explores Gadamer's understanding of practical philosophy and practical
application, but by focussing on a different theme: the firm distinction
Gadamer draws between practical philosophy and *phronesis*, which
Aristotle defined as the intellectual virtue of the ready ability to make

---

2　Hoy, 61.
3　Hoy, 102.

wise moral judgment. The issues raised in this section prove to be particularly significant for the rest of our investigation.[4]

ALTERNATIVE PERCEPTIONS OF HERMENEUTICS AND HUMAN SCIENCE

The question of criteria is really a question of relevance. Gadamer certainly wants his philosophy to be relevant to the spectrum of work done in the human sciences. But does the assurance of relevance consist only in the demonstration of criteria? This is the question with which we should confront those, such as Betti and Hirsch, who press Gadamer on the question of criteria.[5] By implicitly equating relevance with criteria, they misperceive the scope in which Gadamer views the relationship between philosophical hermeneutics and human science. But it is still possible—and important—to raise the question of criteria within the scope of relevance which Gadamer perceives, and so begin to frame an

---

4 The approximate meaning of such terms as 'practical philosophy,' 'phronesis,' and 'practice' will be elaborated as we proceed, though in a larger sense that clarification is the task to which this entire work aims to contribute. But a word should be said here about the use of such terms, since one might well ask why we should use them at all, given their arcane, foreign, or seemingly inappropriate character. Why not consistently 'translate' them into more familiar terms if that is possible?

But such a translation, at least initially, is either not possible, or not helpful since it would only constrict and distort realities which sometimes only new words can help us to perceive. This is certainly true of *phronesis*, which, as presented to us by Aristotle, always seems to mean more than 'prudence,' which is how it is usually translated, and is only roughly connoted by 'wisdom in practical matters,' For now the term is irreplaceable.

As for 'practice,' this certainly is a word in our language, although its conventional meanings (eg., as in "practice makes perfect," or "will it work in practice?") are fragmented and limited aspects of what was once meant by *praxis*. Today many writers use this ancient word instead, projecting an air of esoteric sophistication and giving the impression that its use confers some privilege. It would be better, as is attempted here, to rehabilitate the breadth and nobility of 'practice' in popular usage (which is already etymologically so close to *praxis*), and is enough to occupy us without choosing a word that poses a worse stumbling block. 'Practice' is our native word, and it can and should stand for more than it does.

5 See: Emilio Betti, "Hermeneutics as the General Methodology of the *Geisteswissenschaften*" (1962), in Joseph Bleicher, trans., *Contemporary Hermeneutics* (London: Routledge & Kegan Paul, 1980), 51-94; and Eric Hirsch, "Appendix II. Gadamer's theory of interpretation," in his *Validity in Interpretation* (New Haven: Yale University Press, 1967), 245-264.

intrinsic rather than extrinsic critique. The arena of that inquiry, however, will be moral practice, rather than textual interpretation. But the
question of criteria and its bearing on the relevance of philosophical
hermeneutics presents parallel issues in both areas. It is useful therefore
to briefly examine one effective (and also representative) critique of
Gadamer's hermeneutics as it bears on textual interpretation in the
methodology of the human sciences. Not only will we later see how
similar issues arise with regard to a hermeneutical perspective on
practice, but the answers we arrive at will have relevance for how we
view the understanding of texts.

Hinman's Critique: A Question of Truth

An essay by Lawrence Hinman poses the problem of criteria which
seems to dog the heels of philosophical hermeneutics. Hinman is
sympathetic to Gadamer's hermeneutics yet is troubled by its apparent
lack of criteria for judgment, and hence its apparent irrelevance to the
conscious, practical work of the human sciences. Hinman poses the issue
in this way:

> Gadamer is faced with a dilemma. Either he is presenting only an analysis of
> what is inevitably involved in all understanding, interpretation, and appli-
> cation, or else he is (perhaps also) presenting an analysis of what is involved
> in all *good* (or valid or justified) understanding, interpretation, and applica-
> tion . . . Insofar as Gadamer constantly presupposes in specific instances that
> some interpretations have in fact a greater claim to truth than others, he pre-
> supposes but does not acknowledge a particular answer to the *quaestio juris*.[6]

In Hinman's view, Gadamer's exclusion of the *quaestio juris*, the
question "about what understanding *ought* to be," is both unjustifiable
for a theory of hermeneutics, and "incompatible with a number of fundamental insights that are presented in *Truth and Method*," and must be
therefore addressed out of "inner necessity." In defense of the latter argument, Hinman invokes Gadamer's own view that the truth of a text
overflows the scope of the author's intent, and that this truth itself must
be pursued.[7]

What is truth, Hinman asks, in Gadamer's hermeneutic perspective?
He sees two related emphases: first, truth is "the structure of 'what is,'"

---

[6] Lawrence Hinman, "Quid Facti or Quid Juris? The Fundamental Ambiguity of
Gadamer's Understanding of Hermeneutics," *Philosophy and Phenomenological
Research* 40 (June 1980): 513-514.

[7] Hinman, 512, 513, 514.

the truth of finitude, the "game" of life we are already playing; second, truth has an event-character.[8] It is in this relation between hermeneutics and truth that Hinman sees Gadamer's basic and problematic ambiguity.

> It is, I think, precisely this understanding of truth which leads to the rejection of the *quaestio juris*: if truth is an event which plays itself out independently of our will, then there can be no question of how we should participate in that event and thus no *quaestio juris* in that sense.[9]

If truth is an event independent of our activity, then it would appear that a hermeneutic self-understanding (a consciousness of historical effects, as Gadamer would say) does not contribute to the "emergence of truth." If, however, hermeneutics "guarantees truth in some unique way," as *Truth and Method* appears designed to argue, "then such a self-understanding is presumably to be recommended . . . This . . . is to say what understanding ought to be." Hinman puts this more sharply in terms of the relationship between Gadamer's claims and the human sciences. If the self-understanding of the human sciences is not essential (on the ground that understanding is inevitably a hermeneutic process), then hermeneutics is inconsequential to the execution of the human sciences. But if this self-understanding is indeed essential, and if Gadamer differs with the predominant self-understanding of the human sciences, then Gadamer is in fact addressing the *quaestio juris*.[10]

Gadamer could avoid directly responding to the *quaestio juris*, Hinman suggests, by claiming that the only valid understanding is understanding derived from a self-aware hermeneutic perspective. But this still begs the question as to what counts as valid understanding. Perhaps, then, by shifting from general to particular criteria we can find some clue. For example, Gadamer refers to the idea of the classical and to authoritative tradition in ways which suggest that they constitute criteria. But again, neither is defined by specific content. Gadamer is unable to confirm "particular events . . . as instances of these general concepts," and their circularity becomes clear. Gadamer appears unable to secure the relation between the truth of tradition as the whole of our inescapable reality, and the truth of any particular instance of tradition.[11]

As Hinman points out, Gadamer does seem to imply that the validity of a truth claim is to be determined by its productivity in the larger

---

8   Hinman, 523-525.
9   Hinman, 530.
10   Hinman, 529, 531.
11   Hinman, 531-533.

"game" of tradition.[12] But again this puts the *quaestio juris* outside the scope of any individual judgment or will.[13] Nor, we might add, does Gadamer indicate the criteria of productivity, except in the most general terms, such as that finitude itself be recognized as truth, that the "game" be confirmed to be our true situation. In these circumstances it is easy to see how Hinman arrives at the impression that, for Gadamer, truth "becomes something which we cannot determine: it determines us." If this is so, "hermeneutics consigns itself to an odd kind of irrelevance," since hermeneutics would teach that hermeneutics itself "makes little difference" in the history of human understanding. Finally, Hinman makes clear his own hope and expectation of an adequate philosophical hermeneutic. It is a hope we share, though we still regard his critique as an invitation, to Gadamer and to us his readers, to see what hermeneutics has yet to tell us.

> If Gadamer's understanding of hermeneutics is to escape the resignation implicit in its play model of understanding and truth, and if it is to avoid undermining its own distinctive claim to truth, it must begin with a recognition of the necessity of asking the *quaestio juris* as the question of what understanding, interpretation, and application ought to be. To fail to do this is, in Heideggerian terms, to reduce Dasein to facticity and to ignore the fact that Dasein is in each case the kind of being for whom its own being is an issue. The question of how we understand is for us always simultaneously the question of how we ought to understand insofar as we are concerned with the question of truth.[14]

Gadamer's Reply: A Counter-question of Practice

Gadamer offers what seem to be conflicting responses to the issues facing philosophical hermeneutics which Hinman's critique raises. In the first place, he often denies that his theory has normative intentions or implications. Gadamer responded to a variety of critics when he wrote the "Forward to the Second Edition" of *Truth and Method* in 1965. It disappoints expectations of a positive, methodological answer, and instead leaves us with these often-quoted confounding lines:

> I did not intend to produce a manual for guiding understanding in the manner of the earlier hermeneutics. I did not wish to elaborate a system of rules to describe, let alone direct, the methodical procedure of the human sciences. Nor was it my aim to investigate the theoretical foundation of work in these fields in order to put my findings to practical ends. If there is

---

12 Hinman, 534n cites *TM*, 1st ed., [247, 251, 265].
13 Hinman, 534-535.

any practical consequence of the present investigation, it certainly has nothing to do with an unscientific "commitment"; instead, it is concerned with the "scientific" integrity of acknowledging the commitment involved in all understanding. My real concern was and is philosophic: not what we do or what we ought to do, but what happens to us over and above our wanting and doing.[15]

If Gadamer does not offer a hermeneutical art or technique as an alternative to the methodologism he criticized, does this mean that methodology is not an important question at all? Gadamer replies in the negative:

I did not remotely intend to deny the necessity of methodical work within the human sciences (Geisteswissenschaften) . . . The difference that confronts us is not in the method but in the objectives of knowledge. The question I have asked seeks to discover and bring into consciousness something that methodological dispute serves only to conceal and neglect, something that does not so much confine or limit modern science as precede it and make it possible.[16]

So while methodology is an important question, it is not a subject of his philosophical hermeneutics.

This seems to be Gadamer's meaning when he says "it seems to me a mere misunderstanding to invoke the famous Kantian distinction between quaestio juris and quaestio facti." He goes on to say that Kant's inquiry reflected a philosophical question, not a methodological one. But while Kant's inquiry dealt with the conditions and scope of scientific knowledge, Gadamer sees his own philosophical question as posed not "only of science and its modes of experience, but of all human experience of the world and human living. It asks (to put it in Kantian terms): how is understanding possible?"[17]

Hinman recognizes that Gadamer rejects the applicability to his philosophy of the Kantian distinction between the *quaestio juris* and the *quaestio facti*. Hinman interprets Gadamer's position to be, first, that his hermeneutics does pose the *quaestio juris* in the sense of asking *how* the understanding of human science is possible; and second, that he is not posing the *quaestio juris* in the sense of judging what that understanding *ought* to be. But Gadamer's philosophy, with respect to the first claim, is not likely to find wide agreement in contemporary human science dis-

---

14 Hinman, 535.
15 *TM*, xxviii [xvi].
16 *TM*, xxix [xvii].
17 *TM*, xxix, xxx [xvii, xviii].

ciplines, and Gadamer is obviously concerned with identifying this difference in viewpoint. Thus, contrary to the second claim, Gadamer does in fact appear to be telling the human sciences, in some sense, how they ought to execute their tasks.[18] Hinman appears to be correct here. And if Gadamer is trying to distinguish a methodological inquiry from a philosophical inquiry in this subject, he remains vague about where the difference lies, which suggests to us that such a 'border' cannot remain inviolable.

Fortunately, we are able to pursue this inquiry through a different and very important theme in Gadamer's work. This theme concerns Aristotle's concept of practical philosophy and its contemporary relevance as a hermeneutically oriented model for the human sciences. It constitutes, in effect, a constructive reply to the kind of question raised by Hinman, but it also serves to bring the whole realm of practice into the ambit of philosophical hermeneutics. Nearly all of Gadamer's discussions of practical philosophy appear in essays which constitute or follow Gadamer's 'debate' with Habermas in the late 1960's. Frederick Lawrence's perception seems correct:

> I suspect that Gadamer has been compelled by the challenge to his rather recently influential hermeneutic philosophy on the part of the critical theorists of the Frankfurt school to a more explicit realization of the practical-political implication of his undertaking and especially of the importance for him of Aristotle.[19]

But it is perhaps more important to note, as Lawrence immediately adds, that "it is certain that this provocation initially did not bring about his [Gadamer's] discovery of this dimension of his hermeneutic philosophy."[20] For the roots of Gadamer's inquiry in this direction lie in his pursuit of Aristotle's relevance for hermeneutics, which predates this debate. Gadamer says in *Truth and Method* that "Aristotelian ethics quite unexpectedly made it easier to understand the hermeneutical problem more deeply."[21] The origins of this inquiry by Gadamer go back at least as far as a seminar on Aristotle given by Heidegger in 1923, by which Gadamer was first brought into serious study of Aristotle's ethics.[22] And in a published letter to Richard Bernstein, Gadamer points to even earlier

---

[18] Hinman, 526-528.
[19] Lawrence, "Introduction," in Gadamer, *RAS*, xiv.
[20] Lawrence, "Introduction," in Gadamer, *RAS*, xiv.
[21] *TM*, 540 [490].
[22] See, for example, *RAS*, 47.

origins: "As important as Heidegger and his 1923 *phronesis* interpretation were for me, I was already prepared for it on my own, above all by my earlier reading of Kierkegaard, by the Platonic Socrates, and by the powerful effect of the poet Stephen George on my generation."[23] Robert Sullivan observes that "Gadamer's writings from 1934 to 1942 can well be characterized as a kind of 'political hermeneutics' because of his emphasis on the moral context that sustains the ethical individual." In fact, he says, "philosophical hermeneutics was first of all a different way of doing politics."[24] So while it is true that Gadamer has brought his inquiry into Aristotle's practical philosophy more to the fore in and since his debate with Habermas, that inquiry has been larger than the context of that debate and needs to be treated as an aspect of his whole work. For this reason it is introduced here and later refined through his exchanges with Strauss, Habermas and others.

It should be noted, then, that Gadamer's interest in practical philosophy grew out of its simultaneous relevance for the philosophy of the human sciences and for ethics. Gadamer elaborates on this in his essay "The Heritage of Hegel," which includes several references to his own intellectual odyssey. On the one hand, he says that he learned to understand the existential pathos of Kierkegaard by locating "its prototype in the unity of *ethos* and *logos* that Aristotle thematized as practical philosophy, and especially as the virtue of practical rationality [i.e., *phronesis*]." On the other hand, it was in the course of this inquiry that he found that practical philosophy contains "a moment that offers an exact correspondence to the hermeneutic experience, and especially to that operative in the sciences."[25]

Gadamer's pursuit of the relevance of practical philosophy for human science focusses on the meaning of our technical and scientific civilization and the destiny of philosophy within it:

> We live in an age that would as soon count philosophy among the theological relics of a bygone age or that suspects nothing so much of having a dependence on secret or unconscious interests as the ideal of pure theory and of knowledge for the sake of knowledge alone . . . Since the technological

---

[23] Hans-Georg Gadamer, "Appendix: A Letter by Professor Hans-Georg Gadamer" (1982), trans. James Bohman, in Richard Bernstein, *Beyond Objectivism and Relativism: Science, Hermeneutics and Praxis* (Philadelphia: University of Pennsylvania Press, 1983), 265.

[24] Robert R. Sullivan, "Translator's Introduction," in Gadamer, *Philosophical Apprenticeships*, xvi. See also Sullivan's book, *Political Hermeneutics* (University Park, PA: Pennsylvania State University, 1989).

civilization and the feverish progress with which it has covered the globe
has confronted humanity with breathtaking problems of self-destruction in
war and peace, the passion for philosophy appears altogether like an irresponsible flight into a world of fading dreams. And shall we now assert that
philosophy pertains as essentially to the natural inclinations of humanity as
its technical rationality and its practical shrewdness, the collective impact of
which hardly seems sufficient to cope with humanity's future tasks?[26]

Of course, this is precisely what he wants to assert, and he explains why:
"What the old science, crowned by metaphysics, had provided was a
whole orientation to the world, which brought natural experience of the
world and its linguistically mediated interpretation of the world to a unified conclusion. Modern science could not provide this."[27]

At the same time, Gadamer acknowledges that changed intellectual
circumstances do affect what we can expect of philosophy today. But
through all such changes, "the natural inclination of human beings toward philosophy, toward the desire to know, prevails." This is a desire
to know "that constantly needs to be nourished from other sources than
those of a research making ever further progress." Though modern science "has put an end to the classical function of philosophy," philosophy
has found new avenues for enduring in the humanities, in art, and in that
nineteenth-century phenomenon, the birth of the human sciences. "The
aim of these kinds of science is not just knowledge but the vital and
ongoing shaping of man's knowledge of himself," and this function of
philosophy is irreplaceable for human existence.[28]

This historical perspective on the human function of philosophy
leads Gadamer to articulate a question which is first posed in his exchange with Habermas and reverberates throughout his subsequent
work.

> Where can we find an orientation, a philosophical justification, for a scien-
> tific and critical effort which shares the modern ideal of method and yet
> which does not lose the condition of solidarity with and justification of our
> practical living?[29]

What is Gadamer getting at in this question? He poses it—in two basic
forms—a number of times. In one form, this question is directed at our

---

25   *RAS*, 47, 48.
26   *RAS*, 139.
27   *RAS*, 144.
28   *RAS*, 146-149 passim.
29   "Hermeneutics and Social Science," 311.

understanding of what the human sciences are and how we ought to approach them theoretically. Thus he asks, "How can we develop a concept of knowledge and science which really corresponds with what everyone is doing in the humanities?" Recapitulating *Truth and Method*, he explains that "it is evident that the expression 'human science' is problematic for us today and that we must come to the conclusion that science should be defined by us in another way than it is for modern times." We must face "the necessity of an epistemological self-understanding which is not based on the credence of the natural sciences and of the ideal of method . . . as it dominates the research work and our academic activities in the humanities." At the same time, Gadamer is not endorsing a narrow-minded crusade; he recognizes that it is "a very ambiguous situation in which the humanities" work today. Implicitly referring to his own debate with Habermas, he observes that "It is not very comfortable to choose between the rhetorical tradition of the *artes liberales* or to side for a self-understanding that calls itself critical and methodical and tries to compete with the natural sciences."[30]

In several places, Gadamer also asks his question in another way, one which focusses on modern social and political life. The question quoted in the Introduction to this work is posed in this second fashion: "How can we learn to recover our natural reason and our moral and political prudence? In other words, how can we reintegrate the tremendous power of our technique within a well-balanced order of society and reconstitute a living solidarity?"[31] In another essay Gadamer describes this task as one of "intense urgency" because "we live in a condition of ever-increasing self-estrangement, which far from being caused by the peculiarities of the capitalist economic order alone, is due rather to the dependence of our humanity upon that which we have built around ourselves as our civilization,"[32] which seems to point to our basic dependence on the tools of modern science and technology. Significantly, he elsewhere remarks that the problem of hermeneutics is "not restricted to the areas from which I began in my own investigations," as in *Truth and Method*. "My only concern there was to secure a theoretical basis that

---

[30] Hans-Georg Gadamer, "Practical Philosophy as a Model of the Human Sciences," trans. James Risser, *Research in Phenomenology* 9 (1979): 74-78 passim.

[31] "Hermeneutics and Social Science," 314.

[32] *RAS*, 149.

would enable us to deal with the basic factor of contemporary culture, namely, science and its industrial, technological utilization."[33]

These theoretical and practical aspects of Gadamer's question about a new integrated perspective are linked in two important ways. The first is already implicit: the forms of science involved in both of these aspects share a great deal and must be addressed together.

> We cannot avoid the question of whether what we are aware of in such apparently harmless examples as the aesthetic consciousness and the historical consciousness does not represent a problem that is also present in modern natural science and our technological attitude toward the world . . . Over against the whole of our civilization that is founded on modern science, we must ask if something has not been omitted. If the presuppositions of these possibilities for knowing and making remain half in the dark, cannot the result be that the hand applying this knowledge will be destructive?[34]

The two forms of Gadamer's question are also linked in a second way: in both cases he points to Aristotle's practical philosophy as the model for an answer. With respect to the self-understanding of the human sciences, "Aristotle offers us a better understanding of human life than can modern science"; and with respect to the realm of practice, Gadamer asserts that Aristotle's political philosophy "had a deeper relevance than we imagine."[35]

Gadamer clearly considers the methodological and practical-moral aspects of his question about a new perspective beyond that of modern science to be different branches of the same task, a task which will benefit from inquiry into the significance of Aristotle's practical philosophy:

> In order to work out an orientation which brings together *both* methodological access to our world *and* the conditions of our social life, it was natural for me to return to preceding philosophical orientations and ultimately to the tradition of the practical and political philosophy of Aristotle . . . To justify the procedure proper to this broader field, we must re-enact Aristotle's ideal of *praktike episteme* [i.e., practical science]; this ideal anticipated the crises in method of the modern humanities.[36]

---

[33] Hans-Georg Gadamer, "The Universality of the Hermeneutical Problem" (1966), in his *Philosophical Hermeneutics*, trans. and ed. David Linge (Berkeley: University of California Press, 1976), 10-11.

[34] "The Universality of the Hermeneutical Problem," in *Philosophical Hermeneutics*, 10.

[35] "Practical Philosophy as a Model," 77, 78.

[36] "Hermeneutics and Social Science," 311.

Thus Gadamer's guiding question amounts to an effort to interpret this ideal of practical knowledge and practical philosophy; that is, to apply Aristotle's ideas on this subject to the human sciences of today.[37] Below, Gadamer summarizes the scope of that application of Aristotle in a way which indicates why he regards the theoretical and practical expressions of his inquiry as inextricably linked. This summary also serves, in almost every sentence, to introduce the themes we shall pursue in the following section of this chapter:

> When Aristotle, in the sixth book of the *Nicomachean Ethics*, distinguishes the manner of "practical" knowledge . . . from theoretical and technical knowledge, he expresses, in my opinion, one of the greatest truths by which the Greeks throw light upon "scientific" mystification of modern society of specialization. In addition, the scientific character of practical philosophy is, as far as I can see, the only methodological model for self-understanding of the human sciences if they are to be liberated from the spurious narrowing imposed by the model of the natural sciences. It imparts a scientific justification to the practical reason which sustains all human society and which is linked through millennia to the tradition of rhetoric. Here the hermeneutic problem becomes central; only the concretization of the general imparts to it its specific content.[38]

AFFINITY BETWEEN INTERPRETATION AND PRACTICE

Gadamer's hermeneutic of practical philosophy results in a unique claim, which he expresses several times. As indicated in our Introduction, Gadamer asserts that "hermeneutic philosophy is the heir of the older tradition of practical philosophy."[39] In another essay he asserts "the neighborly affinity of hermeneutics with practical philosophy" in the "most authentic realm of hermeneutic experience," which aims at the "regaining of a shared possession of meaning." He goes so far as to say that "hermeneutics is philosophy, and as philosophy it is practical philosophy."[40] It is quite a broad and challenging claim to link two

---

[37] Elsewhere, Gadamer poses this application in a different way as an inquiry into "how the notable tension in Plato and Aristotle between a technical notion of knowledge and a practical, political notion of knowledge, which includes the ultimate end of human beings, may be made fruitful within the matrix of modern science and theory of science." (*RAS*, 131.)

[38] "Problem of Historical Consciousness," 5.

[39] "Hermeneutics and Social Science," 316.

[40] *RAS*, 109, 111.

streams of thought which are generally considered to be quite distinct. The expectation and potential significance of this claim sustains this inquiry as we proceed.

Gadamer's claim for this affinity is oriented toward both a philosophical conception of the ground of the human sciences, and toward the practical and even political significance of hermeneutics—in the same way as he posed his question about the possible integration of critical technique and communal solidarity. But this double character is to the point: Gadamer really means to cross disciplinary boundaries, to break academic taboos by showing the possibility of a science—a philosophy—of human affairs that is relevant to everything human, both of understanding and of values, both of reflection and of action. Our task here is to gain some sense of what, for Gadamer, legitimates this association of a discipline of understanding with one of valuing, deciding and acting. Of course, Gadamer does not regard hermeneutics and practice as identical disciplines or activities; he wishes to point to basic gestures, perspectives, and self-understandings buried deep inside both of these disciplines and activities. Furthermore, Gadamer is not proposing a modern revival of practical philosophy in its classical form, nor the mutation of hermeneutics into a practical hermeneutics of codified norms for use by either the human sciences or ethics. The point, as developed in this section, is simply to look at what hermeneutics and practice share, at what they have in common that may awaken our thinking to new insights. It is important to remember, however, that Gadamer is primarily concerned to show the practical character of hermeneutical experience, while our ultimate aim is the inverse: to inquire into the hermeneutical character of practice and of philosophy of practice.

The Nature of Practice

Gadamer's hermeneutic of Aristotle begins with the meaning of practice. One such discussion locates the root of practice in the concept of *prohairesis*, which (through translation into German and then into English) is rendered as "free choice," "preference" and "prior choice." This is a characteristic of practice which is unique to human being. But it is located in a broader horizon of practice which we share with animals as well.[41] "Practice, as the character of being alive, stands between activity and situatedness . . . Practice means . . . the actuation of life (*energia*) of anything alive, to which corresponds a life, a way of life, a life that is led

---

[41]   *RAS*, 90, 91.

in a certain way (*bios*)."[42] Human practice, then, "is specified by the *prohairesis* of the *bios*," by the choice of a way of life. For Gadamer, the critical characterization Aristotle makes about practice, even more than its difference from theory, is its distinction from *techne*, the knowledge of making things. This making provides "the economic basis for the life of the polis."[43] Thus the choice of a way of life is most importantly a moral, that is, a practical choice, and not a technical choice. Thus practical philosophy is the 'science' of this human capability of choice, and of which choices are best.

> Practical philosophy, then, has to do not with the learnable crafts and skills, however essential this dimension of human ability too is for the communal life of humanity. Rather it has to do with what is each individual's due as a citizen and what constitutes his *arete* or excellence. Hence practical philosophy needs to raise to the level of reflective awareness the distinctively human trait of having *prohairesis* . . . It has to be accountable with its knowledge for the viewpoint in terms of which one thing is said to be preferred to another: the relationship to the good.[44]

Gadamer recalls for us that Aristotle described the method and work of practical philosophy in such a way that it "corresponds to the special conditions of our practical knowledge." Foremost of these conditions was that "members of society are the only possible students of the rules and constitutional elements of social and political life." And since "morality and politics," the objects of this science, "are not susceptible to a detached theoretical interest," they "presuppose education and maturity."[45]

Gadamer also offers his own thoughts on the basic character of practice. He points, for example, to the "capacity of man to think beyond his own life in the world, to think about death." Such a universal practice as gifts of mourning for the dead reflects "the fundamental constitution of human being from which derives the specific sense of human practice; we are dealing here with a conduct of life that has spiraled out of the order of nature." When life thus transcends nature, choices about ends and means must be made and this results in "the stabilization of norms of conduct in the sense of right and wrong."[46] From consideration of such

---

42 *RAS*, 90.
43 *RAS*, 91.
44 *RAS*, 92.
45 "Hermeneutics and Social Science," 312.
46 *RAS*, 75, 76.

elements of human life, Gadamer urges us to sense the true scope of human practice:

> We have to keep in full view the entire range of the human—from the cult of the dead and concern with what is just, to war—in order to apprehend the true meaning of human practice. It is not exhausted in collective and functional adaptation to the most natural conditions for life, as we can verify among animals that form a state. Human society is organized for the sake of a common order of living, so that each individual knows and acknowledges it as a common one (and even in its breakdowns, in crime). It is precisely the excess beyond what is necessary for the mere preservation of life that distinguishes his action as human action.[47]

Hermeneutical Interpretation and Moral Knowledge

Gadamer's elaboration of the affinity between philosophical hermeneutics and practical philosophy appears in his discussion in *Truth and Method* of Aristotle's argument for the uniqueness of "moral knowledge," that is, knowledge about human practice.[48] Gadamer introduces this topic in the following manner:

> If the heart of the hermeneutical problem is that one and the same tradition must time and again always be understood in a different way [i.e., in each new historical horizon of prejudgment], the problem, logically speaking, concerns the relationship between the universal and the particular. Understanding, then, is a special case of applying something universal to a particular situation. This makes *Aristotelian ethics* of special importance for us . . . [49]

That importance lies in the fact that "if we relate Aristotle's description of the ethical phenomenon and especially the virtue of moral knowledge to our own investigation, we find that his analysis in fact offers a kind of *model of the problems of hermeneutics*."[50] Gadamer says in his 1957 lectures even more specifically that "It is the role of reason and knowledge in Aristotle's *Ethics* which manifests such striking analogies to the role of historical knowledge."[51] The problem which has relevance to hermeneutics is Aristotle's effort to characterize moral (i.e., practical) knowledge as something which, while obviously related to both theoreti-

---

47   *RAS*, 76-77.
48   This topic is discussed at 312-324 [278-289] in *TM*, and is also taken up as part of Gadamer's 1957 lectures on "The Problem of Historical Consciousness" at 30-38.
49   *TM*, 312 [278]. His italics.
50   *TM*, 324 [289].
51   "Problem of Historical Consciousness," 30.

cal knowledge (of unchanging truths) and to technical knowledge (of how to make things), is finally unique and therefore independent of these other categories of knowledge and their methods.[52]

Moral knowledge is not a theoretical knowledge in the sense of unchanging mathematical truths, but the knowledge of an "acting being," who "is concerned with what is not always the same but can also be different. In it he can discover the point at which he has to act. The purpose of his knowledge is to govern his *action*." In this respect, "the human sciences stand closer to moral knowledge" than to that kind of "'theoretical' knowledge. They are 'moral' sciences. Their object is man and what he knows of himself." As Gadamer points out, "Aristotle catches this difference in a bold and unique way when he calls this kind of knowledge, self-knowledge, i.e., knowledge for oneself."[53] But is moral knowledge then a kind of technical knowledge, such as we use in making something? Or is the idea of application in moral knowledge really different than the idea of application in technical knowledge?

If the latter applies, it would clearly serve Gadamer's immediate interest in the nature of moral knowledge: it would provide a way of distinguishing philosophical hermeneutics from a theory of hermeneutics that is based on a technical kind of application of "some pregiven universal to the particular situation."[54] At the same time, Gadamer makes clear that the association between moral and hermeneutical knowledge does not imply that they are the same or share an identical subject matter. With regard to ethical theory:

> It is true that Aristotle is not concerned with the hermeneutical problem and certainly not with its historical dimension, but with the right estimation of the role that reason has to play in moral action. But what interests us here is precisely that he is concerned with reason and with knowledge, not detached from a being that is becoming, but determined by it and determinative of it.[55]

And conversely, with regard to philosophical hermeneutics:

> Admittedly, hermeneutical consciousness is involved neither with technical nor moral knowledge, but these two types of knowledge still include *the same task of application* that we have recognized as the central problem of

---

52 See *TM*, 312-317 [278-283].

53 *TM*, 314, 316 [280, 282].

54 *TM*, 324 [289].

55 *TM*, 312 [278]. The last phrase alludes to the double character of the consciousness of historical effects, discussed in the previous chapter.

hermeneutics. Certainly application does not mean the same thing in each case.[56]

What links hermeneutical, moral and technical problems is that they all concern changing beings and truths, and therefore all involve application, although not of the same kind or in the same field. They correspond to different aspects of the same reality, i.e., the historicity of human being.

Gadamer's concern here is to substantiate why and how moral knowledge is different from technical knowledge, a difference which will on some level also characterize the hermeneutical process, and thereby help identify it and distinguish it from the technical process to which this age tends to reduce hermeneutics. To fulfill this aim, Gadamer directs attention to three specific differences between technical and hermeneutical experience which he develops from Aristotle's discussion of *phronesis* in the *Nicomachean Ethics*. Gadamer's use of this source confirms that when he uses the term *sittlichen Wissen*, which is translated as "moral knowledge" in both English editions, he is referring to this virtue.[57]

The first point Gadamer draws from Aristotle is that while both moral application and technical application concern temporal things at the point of judgment and action, they nevertheless show themselves to be different kinds of knowledge.

> We learn a techne and can also forget it. But we do not learn moral knowledge, nor can we forget it. We do not stand over against it, as if it were something that we can acquire or not, as we can choose to acquire an objective skill, a techne. Rather, we are always already in the situation of having to act . . . and hence we must already possess and be able to apply moral knowledge. That is why the concept of application is highly problematical. For we can only apply something that we already have; but we do not possess moral knowledge in such a way that we already have it and then apply it to specific situations . . . What is right, for example, cannot be fully determined independently of the situation that requires a right action from me, whereas the eidos of what a craftsman wants to make is fully determined by the use for which it is intended.[58]

---

56 *TM*, 281.

57 The correlation is made explicit on 314 of the 2nd edition: ". . . moral knowledge (phronesis) and theoretical knowledge (episteme) . . ."

58 *TM*, 317 [287]. Paul Schuchman puts the contrast between moral and technical knowledge this way: "What is right cannot be wholly determined independently of the situation that requires right action from me, whereas the form of what a craftsman desires to make is already fully determined independently of any particular in-

To illustrate his point that even to 'have' moral knowledge requires a decision within a particular situation, Gadamer compares cases where an apparently incorrect application is made in craftsmanship and in jurisprudence. If the artisan lacks a tool or resource he may decide to complete an 'imperfect' work, i.e., improvise to create something less than what was originally intended or preferred.

> But the fact that he gives up and is content with an imperfect work does not imply that his knowledge of things is augmented or has become more nearly perfect through the experience of failure. On the other hand, when we "apply" a law the situation is entirely different. It can happen that, owing to the characteristics of a concrete situation, we may be obligated to mitigate the severity of the law—but "mitigation" is not exclusive of "application." Mitigation does not ignore the right expressed in the law, no more than it condones an unjustifiable carelessness in its application. When we mitigate the law we do not abandon it; on the contrary, without this mitigation there would really be no justice.[59]

Moral application, then, is always more than technically subordinating a case to its appropriate rule in an automatic fashion; moral application, as illustrated by legal judgment, requires a discerning wisdom about the various 'rules,' written and unwritten, which are appropriate to the particular case.

Gadamer's second point in this discussion of Aristotle is that there is a "fundamental modification of the conceptual relation between means and end, one that distinguishes moral from technical knowledge."[60] To wit, neither the end nor the means to that end can be determined outside of the concrete situation, while at the same time each concrete end participates in the ultimate end of all practice. "The end of ethical [i.e., moral] knowledge is not a 'particular thing,' but . . . it determines the *complete* ethical rectitude of a lifetime."[61] Nor is moral knowledge a second choice substitute when technical knowledge is unavailable. Rather, "moral knowledge always requires . . . deliberation with oneself (*eubolia*) and not knowledge in the manner of a *techne*."[62] The uncertainty of moral decision is not occasional nor is it evidence of a defect; the moral life

---

stance of what is made." (Paul Schuchman, "Aristotle's Phronesis and Gadamer's Hermeneutics," *Philosophy Today* 23 [1979]: 45.)

59 "Problem of Historical Consciousness," 34.
60 *TM*, 320 [286].
61 "Problem of Historical Consciousness," 36.
62 "Problem of Historical Consciousness," 37, 38.

offers no certainties and therefore cannot be addressed by technical knowledge.

> Aristotle's definitions of phronesis have a marked uncertainty about them, in that this knowledge is sometimes related more to the end, and sometimes more to the means to the end. In fact this means that the end towards which our life as a whole tends . . . cannot be the object of a knowledge that can be taught.[63]

The unity which characterizes ends and means is itself imbedded in the purpose of life as a whole. This points to the fact that moral knowledge is more accurately a kind of experience, and significantly, is a primordial rather than derivative kind of experience.

> Moral knowledge is really a knowledge of a special kind. In a curious way it embraces both means and end, and hence differs from technical knowledge. That is why it is pointless to distinguish here between knowledge and experience, as can be done in the case of a techne. For moral knowledge contains a kind of experience in itself, and in fact . . . this is perhaps the fundamental form of experience . . . compared with which all other experience represents an alienation, not to say a denaturing.[64]

Gadamer says this "knowledge of the particular situation . . . is a necessary supplement to moral knowledge;" this is not a sensual perception but a "seeing" which means intelligence, or *nous*.[65] To make explicit what Gadamer implies: human science must try to 'see' in this fundamental way as well, and not be confined only to the limited vision determined by a particular set of rules or methods.

Gadamer's third point is that the knowledge of oneself which constitutes moral knowledge leads to an orientation towards others of which technical knowledge is ignorant. This orientation toward others refers to the "varieties of phronesis," or what in some discussions of Aristotle are referred to as the minor intellectual virtues. Gadamer here refers primarily to *sunesis*, usually rendered as 'understanding,' but here meaning a sympathetic understanding of another person.

> "Being understanding" is introduced [by Aristotle] as a modification of moral knowledge since in this case it is not I who must act. Accordingly synesis means simply the capacity for moral judgment . . . The question here, then, is not about knowledge in general but its concretion at a particular moment. This knowledge also is not in any sense technical knowledge or

---

63   *TM*, 321 [287].
64   *TM*, 322 [287-288].

the application of such . . . Once again we discover that the person who is understanding does not know and judge as one who stands apart and unaffected but rather he thinks along with the other from the perspective of a specific bond of belonging, as if he too were affected.[66]

The non-technical character of *phronesis* is made evident again by the way Aristotle contrasts it with the *deinos*, one who has all the knowledge and shrewdness of a *phronimos*, but uses it for evil purposes and selfish gain. If moral knowledge were simply a form of technical knowledge we would have to regard the *deinos* as a legitimate embodiment of that knowledge. Since this is obviously not true, the *deinos* represents instead a "degenerative form of *phronesis*."[67]

Significance of the Affinity

Gadamer's objective in this analysis is primarily to support his contention that when interpretation has the character of application, it is really more like moral reasoning than the technical implementation of pre-existing rules.

> We, too [like Aristotle], determined that application is neither a subsequent nor merely an occasional part of the phenomenon of understanding, but codetermines it as a whole from the beginning . . . The interpreter seeks no more than to understand this universal, the text—i.e., to understand what it says, what constitutes the text's possessive meaning and significance. In order to understand that, he must not try to disregard himself and his particular hermeneutical situation. He must relate the text to this situation if he wants to understand at all.[68]

However, from another essay it is clear that he also values the affinity between the ends of interpretation and moral practice which his analysis implies: even though their mode is different, the enlargement of a common horizon is the shared aim in both interpretion and practice. "Understanding is an adventure and, like any other adventure, is dangerous," but "affords unique opportunities as well." For "where it is successful, understanding means a growth in inner awareness," a "broadening of our human experiences, our self-knowledge, and our

---

65   *TM*, 322 [287].

66   *TM*, 322-323 [288]. As Gadamer says elsewhere, understanding means "a spirit of discernment of another's moral situation and to the resulting tolerance or indulgence. Now what is this discernment if not the virtue of knowing how to equitably judge the situation of another?" ("Problem of Historical Consciousness," 37.)

67   "Problem of Historical Consciousness," 37, 38.

68   *TM*, 324 [289].

horizons."[69] To perceive the link between these conceptions of the process (application) and the goal (common reality) of hermeneutics is to grasp the scope of Gadamer's philosophical perspective: the realm of practical action is as much a hermeneutical realm as textual interpretation. Indeed, as Gadamer himself suggests, moral judgment is the original hermeneutical experience.[70] "The problem of application . . . represents not only an essential moment in the hermeneutics of religious texts [for example] but makes the philosophic significance of hermeneutic questions as a whole visible; it [application] is more than a methodological instrument."[71] That significance spreads as far as all the human sciences, and those sciences "may justly be permitted to invoke the model of the practical philosophy that could also be called politics by Aristotle."[72]

Our original context for considering this discussion of moral and hermeneutical philosophy was to see whether, and how, Gadamer achieves his aim of a new integration of hermeneutical and methodical knowledge. And in the quotation just above he alludes to this integration by asserting that application is "more than a methodological instrument." But the discussion just surveyed is hardly a thorough treatment of such an integration; in fact, it is the most simple suggestion of one. Nor is this suggestion one with which Hinman, for example, would be satisfied. But what we have gained is Gadamer's relocation of the problem. The discussion can now proceed in the context of practice and the philosophy of practice, which Gadamer implies is the primordial arena in which a new integration must be sought. Instead of addressing the *quaestio juris*, Gadamer has directed the discussion toward a *quaestio praxeos*.

---

[69] *RAS*, 109, 110.

[70] David Hoy concurs with this reading of Gadamer when he identifies the two points which explain Gadamer's claim for the affinity between philosophical hermeneutics and practical philosophy. First, "both involve general thinking, but the generality is restricted by the need for relevance to practical concerns." Although they refer to different kinds of practice, they "have in common . . . a reflection on the essence of different forms of action." Second, "Gadamer's analysis of the nature of understanding is . . . grounded in Aristotle's ethics," and the legitimation of *phronesis* as a real form of knowledge, a form which emerges more from experience than abstract theorizing. It is this sort of experience Gadamer has in mind when he says that we experience tradition, and that the interpretation of tradition is a hermeneutical experience. (Hoy, 57, 58, 59.)

[71] *RAS*, 129.

[72] *RAS*, 136-137.

On the one hand, we have an enlarged understanding of Gadamer's basic claim that "application is an element of understanding itself."[73] But on the other hand, it is also correct to infer, as Richard Bernstein does, what amounts to the inverse: "Understanding, for Gadamer, is a form of *phronesis*." Practical wisdom is more basic than the scientific knowledge associated with *episteme*, and it is the existential horizon for the hermeneutical consciousness. Bernstein writes that "just as *phronesis* determines what the *phronimos* becomes, Gadamer wants to make a similar claim for all authentic understanding—that it is not detached from the interpreter but becomes constitutive of his or her *praxis*."[74] In fact, we saw that Gadamer himself implies this in speaking of Aristotle's concept of sympathetic understanding: "Understanding is a modification of the virtue of moral knowledge."[75] While Aristotle's (and Gadamer's) use of 'understanding' here refers first of all to understanding between individuals in a practical context, it also applies to the whole of human science insofar as it too attempts to understand the meaning of human life, even at a disciplined distance.

This redefinition of the scope of hermeneutics is the basis on which we will be able to demonstrate the relevance of philosophical hermeneutics for concrete practice, a link which will cause us to perceive that philosophy itself in a new way. Aristotle's philosophy of practice is clearly not only a model for philosophical hermeneutics, a resource by which Gadamer defends his claim about the character of textual interpretation. Beyond this, Gadamer means for hermeneutical philosophy itself to be the scene of contemporary practical philosophy, to be its descendent in a different age, the age of modern science.

Building on the affinity he identified between *phronesis* and interpretation in 1960 in *Truth and Method*, Gadamer indicates in the following passage from a 1978 essay how closely he identifies philosophical hermeneutics and practical philosophy:

> Just as politics as practical philosophy is more than the highest technique, this is true for hermeneutics as well. It has to bring everything knowable by the sciences into the context of mutual agreement in which we ourselves exist. To the extent that hermeneutics brings the contribution of the sciences into this context of mutual agreement that links us with the tradition . . . that is efficacious in our lives, it is not just a repertory of methods . . . but philosophy. It not only accounts for the procedures applied by science but also

---

73 *TM*, xxxii [xx].

74 Bernstein, *Beyond Objectivism and Relativism*, 146.

75 *TM*, 322-323 [279].

gives an account of the questions that are prior to the application of every science, just as did the rhetoric intended by Plato. These are the questions that are determinative for all human knowing and doing, the greatest of questions, that are decisive for human beings as human and their choice of the good.[76]

Here he goes so far as to say that both of these reflective disciplines have a "determinative" bearing on our understanding of what it means to live as humans and thus on our "choice of the good." Thus, Gadamer implicitly asserts a set of relationships among four basic concepts: as *phronesis* and interpretation have affinity, so do practical philosophy and hermeneutical philosophy. As for Gadamer's inquiry into textual and historical hermeneutics, this situation poses a revealing analogy: that consciousness of historical effects is to philosophical hermeneutics as *phronesis* is to practical philosophy. Moreover, in each of these philosophies the activity being examined is itself made part of the process of theoretical reflection.[77]

Into these expanded horizons we must introduce some critical questions. How is *phronesis* related to practical philosophy? The foregoing discussion of "moral knowledge," as a knowledge acquired and applied in an essentially non-technical manner, pays little attention to this matter and instead strongly suggests that no clear boundary exists between them. Simply to have spoken of *phronesis* primarily as a body of knowledge disregards the distinction between theory and practice, and also deemphasizes the character of *phronesis* as a virtue, and indeed as a kind of worldly wit ('street smarts') whose good decisions must necessarily also be timely and effective. Furthermore, another corresponding term for moral knowledge appears in Gadamer's 1957 lectures and is translated as "ethical know-how," which connotes a different orientation. It perhaps suggests more of the wit of good practice, but also emphasizes the affinity between practice and technical application—an affinity which in Gadamer's eyes has strict limits, as he always reminds us.

These different shadings in his rendition of *phronesis* do seem to serve a purpose. They suggest that Gadamer wishes to portray theory and

---

[76]   *RAS*, 137.

[77]   One interesting implication of these relationships (which we will have reason to explore in the following chapters) is that just as a closed question corrupts interpretation, so the character of a *deinos* corrupts his or her practice. Similarly, what Gadamer calls a slanted question correlates with 'slanted practice'—practice which appears to be open, perhaps even intends to be open, but has already sealed its decision on certain issues.

practice in as close a relationship as possible by showing that *phronesis* and practical philosophy constantly extend bridges to each other. But this also leads to vagueness and confusion when we try to gather a coherent account of practical philosophy. No doubt, as Aristotle remarks, ethics is necessarily imprecise, but the question is whether Gadamer's account retains—or even introduces—more imprecision than is necessary.

PRACTICAL PHILOSOPHY AND *PHRONESIS*

However, if in *Truth and Method* Gadamer constructively blurs the distinction between theory and practice, in other sources he imposes it quite emphatically, concerned to show that however closely related a particular theory and a particular field of practice may be, they do not amount to the same thing. This is particularly evident in several essays written since *Truth and Method* in which he addresses the relationship between *phronesis* and practical philosophy with more depth. This relationship is by no means the only important question in pursuing an understanding of good practice, as the variety of topics in Gadamer's treatment of practice indicate. But it is perhaps the most critical one in the sense that so much else follows from how this question is answered. In short, practical philosophy's understanding of its own relation to good practice is perhaps its own most difficult task.

The Relationship of Theorizing to Moral Practice

At the heart of Gadamer's view is a very firm distinction: "Practical philosophy is not the virtue of practical reasonableness," that is, *phronesis*. The science of practice cannot produce goodness or wisdom in concrete practice. Gadamer points out that Aristotle describes three branches of knowledge: *episteme*, the science of unchanging things; *techne*, the science of humanly created things; and the science of human practice (for which, we may note, Aristotle gives no single definitive name, but rather the various inquiries of ethics and politics, which collectively correspond to what can be called 'practical philosophy'). But Gadamer emphatically points out that these three sciences are not directed at the "knowledge of a physician or a craftsman or a politician that is always to be found in application but [at] knowledge about what may be said and taught in general about such knowing." In other words, we must "distinguish the

investigation of these performances [of human practice] from the performances themselves."[78]

Gadamer's firm distinction between practical philosophy and
phronesis is contrasted by an equally emphatic affirmation of the apparently contrary characteristic of this relationship. Practical philosophy is
"involved with the all-embracing problem of the good in human life."
For "as the science of the good in human life, it [practical philosophy]
promotes that good itself."[79]

This is not the only place where we can see this seeming contradiction in his descriptions of the relationship between a science or discipline
and the activity on which it reflects. Ethics and ethical life also reflect this
ambivalent relationship, and even if this may simply be a different
incarnation of the same relationship between practical philosophy and
phronesis, it reinforces Gadamer's point. On the one hand, "it is essential
that the ethical sciences—while they may contribute to the clarification of
the problems of ethical consciousness—never occupy the place properly
belonging to concrete ethical consciousness."[80] Indeed Gadamer understands Aristotle to mean "that ethics is only a theoretical enterprise and
that anything said by way of a theoretic description of the forms of right
living can at best be of little help when it comes to concrete application to
the human experience of life."[81] On the other hand, "philosophizing
about ethical matters should have an ethical relevancy." More specifically, ethical reflection should "not only serve to advance theoretical clarity but must also have some impact on the clarity of decision-making and
on our general orientation in practical and political life."[82] In *Truth and
Method*, we find another reference to this tension between theory and
practice with regard to "philosophical ethics." The impasse between general theory and particular practice "not only makes philosophical ethics a
methodologically difficult problem, but also gives the problem of

---

[78]   *RAS*, 115-117 passim. In Gadamer's view, it is unfortunate that "Aristotle does
not at all reflect upon this distinction," which seems to mean that Aristotle does not
discuss the distinction implied by his analysis. Gadamer suggests that this omission
perhaps has been a cause of subsequent errors, and he criticizes perennial scholarly
"attempts to see . . . practical philosophy . . . [as] nothing more than the exercise of
practical rationality." Consequently, Gadamer laments how "the aspect of practical
philosophy relevant for the theory of science has remained quite obscure."(*RAS*, 115,
117)

[79]   *RAS*, 117, 118.
[80]   "Problem of Historical Consciousness," 31.
[81]   *RAS*, 112.
[82]   "Practical Philosophy as a Model," 80.

method a moral relevance." How such theorizing, which aims to help moral practice, "can be possible is already a moral problem" before it is a logical or epistemological problem.[83]

Gadamer characterizes the relationship between philosophical hermeneutics and specific hermeneutical judgments in similar terms. We have seen evidence of this earlier, but the point warrants reiteration. On the one hand, he disclaims that his philosophy has any intent to offer a method or technique or art of correct interpretation.[84] On the other hand, he indicates that philosophical hermeneutics is not a neutral discipline but committed to the fusion of horizons, which is characterized as the essence of understanding.

The contrast—and even apparent contradiction—between different aspects of the relationship between practical philosophy and *phronesis* poses a difficult problem for any attempt to justify the meaning and productive role of practical philosophy. Gadamer acknowledges that even the necessity of practical philosophy itself can be put in question since "the undistorted character of natural moral consciousness" already claims to know "how to recognize the good and duty with unsurpassable exactitude and the most delicate sensitivity." And even if practical philosophy does or can aid practice and does not exist for itself alone, what legitimates it as a science? Gadamer seems to be keenly aware of these problems: "It is a problem of the utmost difficulty to work out the specific conditions of the scientific quality that holds sway in these areas." The difficulty is a very basic one:

> We appear to have an impasse. On the one hand we have the clear evidence that practical decision-making and reasoning is not dependant merely on generalities but must be concretized by our practical reasoning and application. On the other hand, there is the game of theoretical description.[85]

In short, "practical philosophy needs a unique kind of legitimation."[86] The bridge—or even fusion—between practice and theory of practice which Gadamer seems to presuppose in *Truth and Method* now seems completely dismantled.

How does Gadamer respond to this "impasse"? He recalls for us how Aristotle addresses this dilemma with a metaphor that Gadamer invokes

---

83 *TM*, 313 [279].

84 Hans-Georg Gadamer, *The Idea of the Good in Platonic-Aristotelian Philosophy*, trans. P. Christopher Smith (New Haven: Yale University Press, 1985), 164, 166.

85 "Practical Philosophy as a Model," 83.

86 *RAS*, 117, 118.

more than once: "He tells us that it is like the man who tries to hit the goal as an archer, and Aristotle compares his own function [in trying to discern good practice] with this man."[87] Thus, as Gadamer elaborates, practical philosophy serves the function of helping to identify the target of action, and nothing more:

> He will more easily hit the mark if he has his target in view. Of course this does not mean that the art of archery consists merely of aiming at a target like this. One has to master the art of archery in order to hit the target at all. But to make aiming easier and to make the steadiness of one's shooting more exact and better, the target serves a real function. If one applies the comparison to practical philosophy, then one has to begin with the fact that the acting human being as the one who—in accord with his *ethos*—he is, is guided by his practical reasonableness in making his concrete decisions, and he surely does not depend upon the guidance of a teacher. Even so, it can be a kind of assistance in the conscious avoidance of certain deviations that ethically pragmatic instruction is capable of affording inasmuch as it aids in making present for rational consideration the ultimate purposes of the direction of one's action.[88]

The function of practical philosophy, as Gadamer summarizes it and poses it more positively, is this: "It assists our concrete, practical ability to size things up insofar as it makes it easier to recognize in what direction we must look and to what things we must pay attention." Nevertheless, "practical philosophy by itself can give us no assurance that we will know how to 'hit' what is right. Such knowledge remains the end of practice itself and the virtue of a practical reasonableness."[89] The same determination to keep theory and practice in dynamic tension is similarly asserted in regard to "philosophical ethics:"

> Thus it is essential that philosophical ethics have the right approach, so that it does not usurp the place of moral consciousness and yet does not seek a purely theoretical and "historical" knowledge either but, by outlining phenomena, helps moral consciousness to attain clarity concerning itself."[90]

---

[87] "Practical Philosophy as a Model," 83.

[88] *RAS*, 134-135. Transliteration substituted for Greek. See also parallel passages in "Practical Philosophy as a Model," 83, and *Idea of the Good*, 163-164.

[89] *Idea of the Good*, 164, 166.

[90] *TM*, 313 [279].

Practical Application as the Concretization of Good

But what is this right approach? The above quotations do little in their conceptual terms to resolve the dilemma Gadamer poses; they simply assert that both parts of the dilemma must be maintained. Metaphorical language can perhaps again offer insight that is beyond the capability of conceptual logic. In fact, Gadamer has more to say in both languages about the relationship between practical philosophy and *phronesis*. To close this chapter, we interpretively present what seems to be at the heart of Gadamer's responses to this matter. This core is affirmed, but it will become evident that certain questionable assumptions in Gadamer's view—taken up in detail later—have their inception here.

His 'resolution' of the dilemma is predicated, in part, on the assertion that the good toward which *phronesis* aims is to be distinguished from some absolute good; only the former is of central concern to practical philosophy as it aids practical wisdom. Here Gadamer adopts Aristotle's critique of Plato, for whom all goods were absolute and thus were immutable, metaphysical realities. Following Aristotle, Gadamer asserts that "a science of the 'good in general' . . . is meaningless for practical philosophy."[91] The illustration which Gadamer used in discussion with one of his translators is that we must distinguish between the ontological goodness of snow, as good or perfect in itself, and practically good snow, which always concerns a concrete case, as in whether 'this' snow is good for skiing or not. For practice, the metaphysical goodness of snow is irrelevant.[92] Gadamer summarizes the point this way: "Aristotle isolates practical philosophy [from other forms of knowledge] for the decisive reason that what we find to be good in the theoretical realm—'good' here meaning immutable being—is something quite different from the right thing to do (*to deon*) at which the practical rationality of human beings aims."[93]

If absolute ideas of good are irrelevant to practical judgment, we seem left with more particular, changeable and contingent ideas of good, which we then apply in practice. But are we then bereft of any absolute certainty that these particular goods are indeed good, since we have cut them off from any absolute goods which could justify them? Apparently

---

[91] *Idea of the Good*, 160.
[92] P. Christopher Smith, "Translator's Introduction," in Hans-Georg Gadamer, *Idea of the Good*, xxvii.
[93] *Idea of the Good*, 160.

so. The alternative seems to be to deduce good action, if this is possible, by reference to ideas of good that are not only more and more particular but therefore also increasingly uncertain. Perhaps then we are reduced to a kind of 'decisionism,' in the sense that one would claim that 'what I decide to be good, by and for myself, is good.' But this form of application still presumes a kind of certainty—by 'making' it so rather than finding it so—which true practice does not have, and therefore still has the character of technical application. It appears that decisionism, then, is not the direction of Gadamer's thinking either.

Is there any third alternative to absolutism and decisionism as models of practical application? If not, the very framework of such application appears to be cast in doubt. Even statements to be found in Aristotle's ethics and politics about virtues and happiness, polities and constitutions, whether general or particular, are "also not meant at all to be applied to the concrete case of doing the right thing, in the way that technical rules are applied in the procedure of producing something. The right thing to do . . . is not simply a case or instance of a rule."[94] Perhaps practical philosophy has no distinct function to serve for practical life. What way still remains in which practical philosophy has ethical relevance, or a guiding, or targeting role to play?

To begin to see how Gadamer answers this question, we must examine how he himself uses words. Despite his rejection of 'rule and case' reasoning, in fact, he continues to use that language even when he is referring to genuine practice, suggesting that his use of such language is metaphorical. For example, Gadamer says that, "It is the task of the power of judgment (and not, to repeat, of teaching and learning [i.e., in a technical model]) to recognize in a given situation the applicability of a general rule."[95] Although Gadamer uses 'rule and case' language here, he places the emphasis on distinguishing practical judgment—*phronesis*—from the learning of a technical art, which by definition uses the application of rules to cases. Thus practical application has features similar to technical application but is still different.

The 'like but unlike' character of practical application as compared to technical application can be perceived in the following statement about interpretation, which reiterates the affinity which interpretation has with practice: "Understanding, like action, always remains a risk and never leaves room for the simple application of a general knowledge of rules to

---

94 *Idea of the Good*, 163.

95 Hans-Georg Gadamer, "Theory, Technology, Practice: The Task of the Science of Man," trans. Howard Brotz, *Social Research* 44 (1977): 545.

the statements of texts to be understood."[96] In the following significant passage, Gadamer again but also more forcefully uses the very language of 'rule and case' to show that practice is something really different from such language:

> The function of *phronesis* . . . is . . . the application of more or less vague ideals or virtue and attitudes to the concrete demand of the situation. Moreover, this application can not evolve by mere rules but is something which must be done by the reasoning man himself.
>
> That is my point: This application, this concretization of the general is the universal aspect of hermeneutics. To understand *in concreto* what the text in general says, that is the task of the jurist in applying law, that is the task of the teacher in explaining the message of the Bible; and for it one needs "prudence." In whatever connection, the application of rules can never be done by rules. In this we have just one alternative, to do it correctly or to be stupid. That is that![97]

There are no absolute rules or 'meta-rules' by which we can justify more particular or concrete rules of practice, nor can we certify that a given application was correct. And yet the "concretization of the general"—the application of a universal rule, purpose or principle—is the aim of practice. The abrupt conclusion to the passage reveals a degree of passion not frequently seen in Gadamer's texts. It also suggests another almost paradoxical comparison between *phronesis* and *techne*: as in the situation of the craftsperson who must know and does know precisely how to swiftly apply a rule to a case, so the practical person must also act. One aspect of the difference is that there is only one chance to act well. These exact conditions will never again arise for the practical person, although they might for the person of craft or *techne*. Perhaps more amazing than the scope of evil or failure in the world is how often humans do good, somehow meeting standards of excellence no technician could meet.[98]

---

96 *RAS*, 109.

97 "Practical Philosphy as a Model," 82-83. Along similar lines, see this statement: "Aristotle's main point—and it is also Plato's—is that science, like the *technai*, like any form of skill or craftsmanship, is knowledge that has to be integrated into the good life of the society by means of *phronesis*. The ideal of a political science that is not based on the lived experience of *phronesis* would be sophistic from Aristotle's point of view." ("Gadamer on Strauss," 12.)

98 Another reader of Gadamer, P. Christopher Smith, supports this general interpretation. Putting Gadamer's point into a first person voice, he says:

> For in moral reasoning I always find myself within a particular situation, and the task is not to subsume this particular case under a universal rule which I could know apart from the situation I am in,

'Rule and case' language appears to be an almost unavoidable metaphor for the activity of practical judgment—but still a metaphor, and a notably limited one at that. In this case, the judgment to which application leads is not essentially a logical operation. *Phronesis* simply *knows*—which also means *risks*—the practically good act which is necessary or best, here and now. These thoughts are epitomized in Gadamer's idea that "there are no rules governing the reasonable use of rules."[99] This warning aphorism—which humbles the intellect but dignifies the spirit—is prominently displayed in several of Gadamer's works, often accompanied by references to the corroboration provided by theological homiletics and by the idea of equity in jurisprudence.[100] Gadamer credits this idea to Kant's *Critique of Judgment*,[101] but as we shall continue to see, it also finds a very natural home in Gadamer's philosophy of the hermeneutical character of understanding and practice, and is placed near the center of that philosophy. More than once, he explicitly

---

> but to define from within my situation what the general rule is of which this situation is an instance. The particular virtues and vices in general, as finding the "mean between the extremes," are not universal principles that I apply to a situation, but universalizations of what I am doing when I do what is right. (P. Christopher Smith, "Translator's Introduction," in Gadamer, *Idea of the Good*, xxvii.)

In a quite unrelated commentary on Aristotle we find additional support. In his essay, "Deliberation and Practical Reason," D. Wiggins makes this observation:

> I entertain the unfriendly suspicion that those who feel they *must* seek more than all this provides want a scientific theory of rationality not so much from a passion for science, even where there can be no science, but because they hope and desire, by some conceptual alchemy, to turn such a theory into a regulative or normative discipline, or into a system of rules by which to spare themselves some of the agony of thinking and all the torment of feeling and understanding that is actually involved in reasoned deliberation. (in *Practical Reasoning*, ed. Joseph Raz [Oxford University Press, 1978], 150.)

Deliberative agony may or may not be present; after all, the point of cultivating *phronesis* as a habit is to allow us to move past those points where others may be diverted, frightened or confused. But it is certain that there is no dearth of ways by which we, in every life situation, seek to reduce practical application to technical application because it is 'easier.'

[99] *RAS*, 121.

[100] See, for example, *RAS*, 49: "The universality of the rule is in need of application and for the application of rules there exists in turn no rule."

[101] See *RAS*, 49, 121.

identifies the concretization of the general as being fundamental to both hermeneutical understanding and practical judgment. "In the end it was the great theme of the concretization of the universal that I learned to think of as the basic experience of hermeneutics."[102] As interpretation aims at a concretization of understanding in a fusion of horizons, so practice aims at concretizing the universal in a particular good.

None of this means that practical philosophy merely exists "to enrich theoretical insight for its own sake." The point is "to apply this knowledge in a reasonable way in the given circumstances of a given situation," an application which *phronesis* alone makes.[103] All of Aristotle's "sketchy descriptions of the typical are rather to be understood as oriented toward such a concretization."[104] There is one location where Gadamer joins his approval for the metaphor of the archer to his notion about the absence of meta-rules: "Aristotle avails himself of this splendid image [of the archer] to say that the theoretical instruction that can be given in practical philosophy puts no rules in one's hands that one could follow in order to 'hit' what is right in accordance with an art," that is, through a technical knowledge.[105]

In fact, Gadamer goes to some lengths to exclude certain features which the modern mind might expect to associate with a theory designed to aide us in doing good practice:

> It is not confined to a particular field. It is not at all the application of a capability to an object. It can work out methods—they are more like rules of thumb than methods—and it can be elevated like an art that one possesses to the stage of genuine mastery. In spite of these things, it is not really 'know-how' which like some knowing-how-to-make just chooses its task (on its own or on request); [rather,] it is posed precisely in the way that the practice of one's living poses it.[106]

What Gadamer is referring to here, what he is defining by exclusion, is essentially his concept of application. A primary purpose of practical philosophy appears to be that of telling us how we ought to—or how we might best—apply theory in practice. The creation of practical good occurs in the course of application. And as often seems the case in such things, Gadamer has an easier time saying what this is not, rather than

---

102 *RAS*, 49.
103 "Practical Philosophy as a Model," 84.
104 *RAS*, 133-134.
105 *Idea of the Good*, 164.
106 *RAS*, 135.

what it is. Practical philosophy is not something we can impose on practice like a grid; it is not an ultimate authority; it does not apply itself. *We* apply it, which means that we exercise *phronesis* to receive something conveyed by practical philosophy, to seize—or be seized by—something to which it points, and to make it our own.

In retrospect, Hinman's question about the criteria of truth appears foreign to the perspective Gadamer proposes: application understood as a tool for making is unlike application understood as a statement in a conversation. What this hermeneutical concept of application means— with regard both to theorizing for the sake of good practice, and to simply understanding practice itself—requires further substantiation and testing, and this is the underlying aim of the following two chapters. All of this puts practical philosophy in a very unique relationship to its own object. We cannot define that relationship by what is said here, nor adequately resolve the dilemma posed earlier. But we can secure the space in which to see events in that relationship as they occur, and perhaps find other ways, as we proceed, to describe them.

Gadamer's implied response to Hinman's question is to enlarge the arena of our attention, which does not neutralize his question but can allow us to re-fashion it. Truth has much more to do with practice than logic alone will permit us to acknowledge. Truth which is significant for humans as humans is often more like the concretizing judgment of practical reason than a deduction from principles. This places Hinman's question in the context of a wider inquiry, and makes Gadamer's "ambiguity" about the criteria of truth something less perplexing than it has been for some readers. By enlarging the horizon of that question so that it includes Gadamer's analysis of the affinity between hermeneutics and *phronesis*, and by inquiring into the nature of practical philosophy and practical application, we are in a position to really begin to ask Gadamer an open question, and not merely confront him with a charge.

THREE

Strauss and Gadamer: Theorizing and Historicity

As indicated at the outset of this investigation, our intent is to investigate the possibility, significance and scope of a 'hermeneutical philosophy of practice,' which we understand to be a philosophy guided by the principles of philosophical hermeneutics, about the nature of practice, and oriented toward the benefit of good practice. This chapter and the following chapter explore the adequacy of the basic theoretical framework of the hermeneutical philosophy of practice in the face of critical objections posed by Leo Strauss and Jürgen Habermas, respectively. The first section of this chapter introduces the underlying issue taken up in both chapters.

TESTING GADAMER'S CLAIMS IN TWO DEBATES

The general objection to the possibility of a hermeneutical philosophy of practice has been the charge that philosophical hermeneutics is a relativist philosophy and therefore has little or nothing to contribute to the understanding of good practice or to the substantive guidance of good practice. If instead hermeneutics does make such normative moral claims, it has contradicted its central claim to simply describe what universally happens in the process of human understanding.

We can refine this problem by posing it in terms of two different but related objections. The first is that philosophical hermeneutics is *inher-*

79

ently relativist, that is, that the premise or conclusion (depending on one's viewpoint) of this philosophy is that there are no universal and absolute moral truths or goods, and that all such claims are illusory because of their inherently historical and contingent character. According to this view, philosophical hermeneutics—regardless of Gadamer's intent—is ultimately nihilistic: it cannot affirm that any good is real. The second form of this charge of relativism is that the philosophy of hermeneutics is *contextually* relativist, that it claims that moral values are true only where their context has determined them to be true. According to this view, philosophical hermeneutics necessarily implies that all moral claims are ultimately prisoners of their own finite horizons, of the social and historical forces which constitute each landscape. In this case, if philosophical hermeneutics does assert moral claims, those claims have no universal validity and cannot be relied on to guide us toward good practice. This again leads ultimately to nihilism.

In either case, then, according to this reasoning, philosophical hermeneutics has no normative power or even relevance for practical philosophy, except possibly an evil one (if on other grounds we still maintain that a nihilistic philosophy is bad). If philosophical hermeneutics is ultimately nihilistic about good, it only abets the forces of modern society which already seem oriented in that direction, and further atrophies any remaining confidence in moral judgment. If it is ultimately ideological itself or merely sustains the ideologies of those who use it, it again abets the breakdown of confidence in the possibility of genuine practice, only more subtly and slowly.

How, therefore, can a philosophy of understanding which takes the contextual, temporal and finite character of understanding as a thesis be relevant to the goal of good practice? If we cannot show that these charges are false—or, if they are true, fail to show that the alleged consequences do not follow—it indeed appears that philosophical hermeneutics is of no substantive use to restoring confidence in the possibility of good practice. The criticisms made and implied by Strauss and Habermas roughly correspond to the two forms of the charge of relativism discussed above: Strauss raises objections to what he sees as the inherent relativism of hermeneutics, and Habermas to its alleged contextual relativism. In this way the abstract objections to the hermeneutical philosophy of practice posed above will be dealt with concretely. In both cases our inquiry benefits from the fact that Gadamer has directly responded to each critic.[1]

The following discussion of the issues raised between Strauss and Gadamer is in four parts: first, Strauss' critique of philosophical hermeneutics, including his arguments against historicism and his challenge to Gadamer contained in correspondence with him; second, an inquiry into Gadamer's ontological perspective, which forms the basis for understanding Gadamer's view of relativism; third, explication of Gadamer's direct and indirect replies to Strauss; and finally, a constructive interpretation of the adequacy of Gadamer's position and of the basis of his differences from Strauss.

STRAUSS' CRITIQUE

The challenge represented by the conversation between Strauss and Gadamer is whether philosophical hermeneutics can rebut the charge that it is a relativist philosophy, and therefore unable to make any truth claims on which practical judgment could depend. This issue focusses attention on the character of truth in an age of historical consciousness and ultimately on the meaning of historicity for human theorizing and self-understanding.

---

1 A few comments will help situate these three writers and our treatment of them. All three authors share a concern to articulate (what amounts to) a practical philosophy, which would be intelligible, true and appropriate to our time. They understand their work as critical of the contemporary, prevailing practical philosophy (in the same generic sense), which they see as corrupted and deformed in some fashion. They are also critical in some way of the present state of social and political affairs; they come to their work with a general judgment that things have gone wrong for reasons which are in part unique to our age and need to be put right.

Beyond these areas of agreement, they differ in many important ways. One reflection of this is that Strauss and Habermas are often associated with the political right and left, respectively. Indeed, it could be said that in some respects Strauss' classical conservatism is situated to Gadamer's 'right' and Habermas' location in the neo-Marxian school of Critical Theory is to Gadamer's 'left.'

It should be noted that the respective projects of Strauss and Habermas are not our major concern. These debates serve our concern primarily as locations in which to examine the theoretical basis of the hermeneutical philosophy of practice and its capacity to withstand the objections represented by their criticisms. Therefore, no attempt is made to adequately present their whole positions, nor, in Habermas' case, the work he has done since his debate with Gadamer.

Arguments Against Historicism

Strauss' arguments against historicism reflect the vision of philosophy as well as of political life which he defends throughout his work. His two major arguments against historicism are best expressed in works which do not refer to Gadamer, although they do include references to Heidegger, Gadamer's teacher and colleague, as an historicist. Whether Gadamer's theory is in fact a suitable object of such objections is, of course, part of the issue at hand; it will be clear, however, why Strauss might well have considered this to be the case. The first argument is that historicism cannot dissolve the legitimacy of political philosophy, as it claims to do, because such reflection on politics is rooted in philosophy itself, which shows that there are unavoidable fundamental questions, the answers to which, Strauss asserts, are in principle accessible even if philosophy never actually acquires final answers.

Strauss presents this argument in the form of rebuttals to two common historicist claims. From the plurality of political philosophies in history, historicism infers that these philosophies refute each other, and therefore concludes that any universal or absolute political or moral truth is impossible. Strauss counters, however, that history "teaches us merely that they contradict each other." It is a philosophical question to then determine which of two of more contradictory theses is true. The fact that this determination has not yet been universally agreed upon is not a refutation of political philosophy but evidence of the inherent character of philosophy. "What else would this mean except that we do not truly know the nature of political things nor the best, or just, political order? This is so far from being a new insight due to historicism that it is implied in the very name 'philosophy.'"[2]

Historicism also asserts that each political philosophy is so clearly related to its historical context that its validity, if it has any, is limited to that context. Strauss replies that such criticism does not recognize how political philosophers have always engaged in "deliberate adaptation" of what they considered to be the absolute political truth to the contingencies of their immediate political situation. They were, in effect, engaged in two expositions, of the absolute truth and of "what they considered desirable or feasible in the circumstances." Thus, the contextual character of political philosophies reflects this "adaptation" of absolute truth, and "does not at all prove that no doctrine can simply be true." In fact,

---

[2] Leo Strauss, "Political Philosophy and History" (1949), in *What is Political Philosophy?* (New York: Free Press, 1959), 62.

Strauss asserts, some settings may be "particularly favorable to the discovery of *the* truth," and this truth would not be refuted simply when the historical setting of its articulation ceases to exist.[3]

With these rebuttals to historicist claims, Strauss indicates that he regards political and moral truth as something which is immutable instead of changeable. Political philosophy is seen as the search for *"the* truth", which will appear, or has appeared, to our consciousness in some particular historical setting. Strauss makes this explicit when he asserts that history shows all thought to have been concerned with the same fundamental themes and problems, and positively infers from this that the same "unchanging framework persists in all changes of human knowledge of both facts and principles." From the human capacity to be aware of this framework he further concludes that human thought is capable of "transcending its historical limitations or of grasping something transhistorical." And this would be true even if all efforts to solve such philosophical problems were shown to necessarily fail "on account of the 'historicity' of 'all' human thought."[4] Consequently, the premise of natural right theory is correct: "the fundamentals of justice are, in principle, accessible to man as man." This conviction rests finally on "the evidence of those single experiences regarding right or wrong," and which "historicism either ignores or else distorts." At the same time, knowing what is natural right remains a problem, and this is why the study of natural right is part of philosophy: "philosophy is knowledge that one does not know, or awareness of the fundamental problems and, therewith, of the fundamental alternatives regarding their solution."[5] Here Strauss and Gadamer seem to agree on the finitude of human knowing. But for Strauss this is no reason to think that natural right (or any political truth) does not have an absolute reality independent of our awareness or finite understanding of it. Gadamer, on the other hand, seems to imply that when horizons fuse, we are witnessing the only kind of truth there can be.

The independent reality of truth is also the premise of Strauss' second argument: historicism destroys itself when its thesis—that all knowledge is historically contingent—is applied to itself, an application it must make in order to be consistent. Historicism "merely replaced one

---

[3] Strauss, "Political Philosophy and History," 63, 64. Italics his.

[4] Leo Strauss, "Natural Right and the Historical Approach" (1953), in Hilail Gildin, ed., *Political Philosophy: Six Essays by Leo Strauss* (Indianapolis, IN: Bobbs-Merrill, 1975), 145, 146.

[5] Strauss, "Natural Right and the Historical Approach," 150, 153, 154.

kind of finality by another kind of finality, by the final conviction that all answers are essentially and radically 'historical.'" But this finality is implicitly disallowed by historicism itself. Historicism could remain internally consistent only "if it presented the historicist thesis not as simply true, but as true for the time being only." Since "historicism must be applied to itself" in this way, showing itself to be "relative to modern man," it proves to be something which "will be replaced, in due time, by a position which is no longer historicist."[6] Strauss' second argument is summarized in this passage:

> The historicist thesis is then exposed to a very obvious difficulty which cannot be solved but only evaded or obscured by considerations of a more subtle character. Historicism asserts that all human thoughts or beliefs are historical, and hence deservedly destined to perish; but historicism itself is a human thought; hence historicism can be of only temporary validity, or it cannot be simply true. To assert the historicist thesis means to doubt it and thus to transcend it. As a matter of fact, historicism claims to have brought to light a truth which has come to stay, a truth valid for all thought, for all time: however much thought has changed and will change, it will always remain historical . . . Historicism thrives on the fact that it consistently exempts itself from its own verdict about all human thought. The historicist thesis is self-contradictory or absurd.[7]

Strauss comments on historicism's response to this logical contradiction. Historicists claim that their thesis reflects an experience of historicity, an experience which "cannot be destroyed by the inevitable logical difficulties from which all expressions of such experiences suffer." While historicism asserts the universal validity of its thesis, it does not assert the universal possibility of its thesis: the historicist insight was not possible in every age, but is only the fate of the modern age. "It is due to fate that the essential dependence of thought on fate is realized now, and was not realized in earlier times . . . The self-contradictory character of the historicist thesis should be charged not to historicism but to reality." In Strauss' view this explanation is merely a rationalization for not facing the absurdity of the historicist thesis. Furthermore, when historicists view the prospect of the disappearance of historicism as a "decline," they are ascribing an absoluteness to their doctrine which the doctrine itself does not allow.[8]

---

6   Strauss, "Political Philosophy and History," 72, 73.
7   Strauss, "Natural Right and the Historical Approach," 146-147.
8   Strauss, "Natural Right and the Historical Approach," 149, 150, 154.

Letters to Gadamer

Strauss' criticisms of historicism pose issues which Gadamer must address. Does philosophical hermeneutics imply the impossibility of independent or universal philosophical truth? If so, does it imply the kind of self-contradiction which would refute such a relativism? Both of these questions are implicitly posed in two letters which Strauss wrote to Gadamer in 1961 after reading *Truth and Method*. Thus, Strauss makes clear that he thinks Gadamer is a proper target of the objections we have just surveyed. This personal challenge between friends and professional colleagues also brings to the fore the existential significance of the theoretical debate about historicism and relativism.[9] This response by Strauss to *Truth and Method* involves a surprising—and even ironic—effort to make sense of Gadamer's hermeneutics in a way which does not lead to the nihilism which Strauss associates with relativism.[10]

---

[9]   Gadamer provides some brief information in two sources regarding his relationship with Leo Strauss. They first met in about 1920 (when both of them would have been about 20 years of age), while Gadamer was a student at Marburg University (see "Gadamer on Strauss," 1, and *Philosophical Apprenticeships*, trans. Robert Sullivan [Cambridge, MA: MIT Press, 1985], 7-20). "We were on good terms and talked now and then but otherwise had few relations with each other." He says their "first real acquaintance" came during Gadamer's 1933 trip to Paris (his first trip outside Germany): he says they spent "a very pleasant ten days together," but also talked about the changing political situation in Germany.("Gadamer on Strauss," 1, 2.) In his *Philosophical Apprenticeships*, 74, Gadamer simply says that in Paris he "often saw" Strauss, who, he points out, had by then permanently left Nazi Germany (presumably, we may add, because he was Jewish), and was working on his book on Hobbes (probably his 1936 work, *The Political Philosophy of Hobbes*).
     Sometime before World War II, Gadamer wrote an article on Plato and Aristotle and says he sent a copy to Strauss. "He wrote me a letter (destroyed during the war) in which he praised it but objected to my using certain modern terms, such as 'sedimentation,' to elucidate Aristotle's thought. That was exactly the point on which we disagreed. To go into the meaning of a text does not require us to speak its language. One can speak the language of another epoch." ("Gadamer on Strauss," 3.)
     They did not meet again until 1954, when, at Gadamer's invitation, Strauss gave a lecture at Heidelberg on Socrates. Gadamer says that Strauss would often refer, in later years, to his conversations with Gadamer in 1954, calling them (in Gadamer's words) "one of the most profitable conversations" of a long time. ("Gadamer on Strauss," 2, 5.) There were other subsequent encounters in the United States. Gadamer does not describe these, although he acknowledges the gracious invitations and receptions he says he received from American followers of Strauss, due, he felt, to Strauss' positive personal references to his professional peer. ("Gadamer on Strauss," 5, and *Philosophical Apprenticeships*, 74.)
[10]   Leo Strauss and Hans-Georg Gadamer, "Correspondence Concerning *Wahrheit und Methode*," *Independent Journal of Philosophy* 2 (1978): 5-12. Five years after Strauss'

In his letters, Strauss expresses his concern about the historicist character of philosophical hermeneutics. "Against your will you seem to preserve 'the historical consciousness of universal comprehension.'" But "how is this possible given the 'finiteness' of man," which is axiomatic for Gadamer himself?[11] Referring to specific passages in *Wahrheit und Methode*, Strauss points to the logical problem of relativism which Gadamer's historicism shares:

> The most comprehensive question which you discuss is indicated by the term "relativism." You take "the relativity of all human values" (54), of all world-views (423) for granted. You realize that this "relativistic" thesis is itself meant to be "absolutely and unconditionally true" (424). It is not clear to me whether you regard the "logical" difficulty as irrelevant (which I would not) or as not by itself decisive.[12]

But Strauss also implicitly poses his first argument when he expresses doubt about whether Gadamer has taken seriously enough the modern crisis of confidence in moral and political truths. He calls *Truth and Method* "the most important work written by a Heideggerian," which presents a "comprehensive doctrine" in the form of a translation of Heidegger into a more academic medium, which focusses on the methodical rather than substantive character of a universal hermeneutics. But perhaps, Strauss implies, this methodical focus is too exclusive.

> It does not appear from your presentation that the radicalization and universalization of hermeneutics is essentially contemporary with the approach of the 'world-night' or the *Untergang des Abendlands*: the 'existential' meaning of that universalization, the catastrophic context to which it belongs, this does not come out.[13]

Strauss sees Gadamer as following in Heidegger's path, and wonders why Gadamer has not addressed the existential significance of hermeneutics in the way Heidegger does—as a kind of refuge for meaning while the world passes through a 'world-night' of disbelief in re-

---

death, three letters exchanged by these authors were published with Gadamer's permission. They consist of a letter by Strauss, dated February 26, 1961; a reply by Gadamer dated April 5, 1961; and a second and final letter by Strauss on May 14, 1961. Gadamer's German letter and an English translation by George Elliot Tucker both appear. Strauss' letters were both in English. The editor indicates that "no further correspondence on these subjects took place." (5)

11   "Correspondence," 5.
12   "Correspondence," 7.
13   "Correspondence," 5.

ceived traditions. However, Strauss holds Gadamer to this Heideggerian theme not only because it is part of Gadamer's own philosophical background but, as several passages make clear, because Strauss himself perceives the present as a 'world-night' and to this extent finds Heidegger congenial to this own thinking.

This is the background to the following, rather surprising statement by Strauss after he challenges Gadamer to address the logical difficulty of his philosophy's relativism:

> I believe there is no "logical" difficulty for the following reason. The historical situation to which the universal hermeneutic or the hermeneutic ontology belongs is not a situation like other situations; it is "the absolute moment"—similar to the belonging of Hegel's system to the absolute moment in the historical process. I say similar and not identical. I would speak of a negatively absolute situation: the awakening from *Seinsvergessenheit* [the state of forgetting being] belongs to the *Ershütterund alles Seienden* [shaking of everything that exists], and what one awakens to is not the final truth in the form of a system but rather a question which can never be fully answered—a level of inquiry and thinking which is meant to be the final level.[14]

It seems clear that Strauss is suggesting how Gadamer might argue that his relativism presents no logical problem by invoking what Strauss sees as a Heideggerian perspective. What is not so clear is whether Strauss himself is also supporting this explanation in some way: first he says there is a problem, then he explains why there is none. Taking the letters as a whole, it appears that Strauss is in some undefined and limited sense supporting this explanation, though perhaps simply to test on Gadamer an idea which is so uncharacteristic of Strauss' other works.

He goes on to interpret Gadamer's project in the same apparently sympathetic light. Existence, he understands Gadamer to say, is itself a process of understanding and therefore of evaluating. "This means that existence is necessarily existence within or through a specific *Sitte-Sittlichkeit* [customary or habitual morality] which is binding, not merely imposed, but as understood, as evident . . . This means that for existence the problem of relativism never arises." The same is then applied to Gadamer's "hermeneutic ontology" (to which we will shortly turn): it is itself rooted in a specific customary morality. The implication is that, for this reason, the problem of relativism does not arise for this ontology.

---

14 "Correspondence," 7. Bracketed translations added.

But it is significant that what stands out for Strauss is the Heideggerian function of such an ontology:

> *The thematic ontology belongs to a world in its decay* when the *Sitte-Sittlichkeit* peculiar to it has lost its evidence or binding power and that therefore the hermeneutic ontology must—of course not dream of fabricating a new *Sitte-Sittlichkeit* but—prepare men for its possible coming or make men receptive to its possible occurrence.[15]

What seems to worry Strauss here is that Gadamer's silence on the existential meaning of his relativism leaves philosophical hermeneutics in the position of abetting the modern perception of the arbitrariness of the universe. In such a vacuum, Strauss invokes Heidegger's hope for the future appearance of a common reality.

It is ironic to see Strauss turn to Heidegger in this way, for Heidegger is frequently and severely criticized on both moral and philosophical grounds in Strauss' other works. In one essay Strauss refers to him anonymously as "the most radical historicist," whose contempt for human "permanencies," such as the distinction between what is base and what is noble, permitted him "in 1933 to submit to, or rather to welcome" the rise of Hitler.[16] To call Heidegger "the hard center of existentialism," which for Strauss is the practical or ethical corollary to historicism, was an epithet aimed at highlighting the absence of any political philosophy in Heidegger's historicist philosophy.[17] The warning to any who might think to continue where Heidegger left off is clear. As to Heidegger's philosophy itself, Strauss perceives the same problem he sees in all historicism. On the one hand, Heidegger asserts that one can only understand a writer creatively and differently than the writer did, not better than he or she did. But on the other hand, in saying that his predecessors were blind to true metaphysics, he is making just such a claim for his own philosophy's superiority.[18] Of course, Strauss is only invoking that part of Heidegger's work which looks to a new tradition after the con-

---

15 "Correspondence," 7. Italics his; bracketed translations added.

16 Leo Strauss, "What is Political Philosophy?" (1959), in Hilail Gildin, ed., *Political Philosophy: Six Essays by Leo Strauss* (Indianapolis, IN: Bobbs-Merrill, 1975), 23, 24. Also see Leo Strauss, "As Unspoken Prologue to a Public Lecture at St. John's" (delivered in 1959), *Interpretation* 7, no. 3 (1978): 2, where he calls Heidegger the "intellectual counterpart to what Hitler was politically."

17 Leo Strauss, "Philosophy as Rigorous Science and Political Philosophy" (1969), *Interpretation* 2, no. 1 (1971): 2.

18 See, for example, Leo Strauss, "Relativism," in Helmut Schoeck, ed., *Relativism and the Study of Man* (Princeton, NJ: Van Nostrand, 1961), 155.

temporary 'world-night.' But even this seems to demonstrate that Strauss is more affected by the perspective and results of historicism than he elsewhere acknowledges. His qualification of Heidegger, and his own convictions about natural law, are apparent when he adds that even in this world-night,

> the basic distinction between the noble and the base and its crucial implica-
> tions . . . retain their evidence or binding power . . . *Above all*, these *things—
> in contradistinction to "world" . . . and other Existentialien—are necessarily the-
> matic within all "horizons."*[19]

For Strauss, this "basic distinction" is an irreducible moral minimum, of which no philosophy can afford to let go, even in this age of historicism, lest we forget, in particular, what happened after 1933. Given the context of Strauss' attempt to respond, both sympathetically and critically, to Gadamer's hermeneutical ontology, this affirmation of the difference between base and noble appears to be an effort to make explicit what Gadamer leaves implicit, and which might distinguish him from Heidegger in Strauss' eyes. The "generality" of distinctions like "noble" and "base," "love" and "hatred," "does not deprive them of definite meaning as you [i.e., Gadamer] yourself make clear."[20]

In summary, Strauss' letter reflects a judgment against Gadamer's historicism, but also a query—reflecting a guarded but wistful hope—about whether philosophical hermeneutics is in fact something other than a relativist philosophy, or is still bound to the nihilistic consequences of modern historicism. The challenge is more explicit in his brief second letter to Gadamer, apparently written after finding Gadamer's initial reply unsatisfactory. He observes that Gadamer's teaching on art and language "is not in any way a traditional teaching," that is, it is not pre-historicist, and therefore, Strauss infers, must be motivated by a Heideggerian concern. Strauss puts the question to Gadamer in this way:

> It is necessary to reflect on the situation which demands the new
> hermeneutics, i.e. on our situation; this reflection will necessarily bring to
> light a radical crisis, an unprecedented crisis and this is what Heidegger
> means by the approach of the world night. Or do you deny the necessity
> and the possibility of such a reflection? I see a connection between your si-

---

19 "Correspondence," 7. Italics his.
20 "Correspondence," 7.

lence on this crucial question and your failure to reply to my remarks regarding relativism.[21]

In short, Strauss asserts, "I believe that you will have to admit that there
is a fundamental difference between your post-historicist hermeneutics
and pre-historicist (traditional) hermeneutics."[22]

It is apparent from his letters that Strauss himself is not quite sure
how to deal with historicism, given the modern loss of confidence in received traditions. He even appears to hope that somewhere in the
hermeneutical perspectives of Heidegger or Gadamer he will find some
clue which will prevent the nihilistic consequences he fears and instead
provide a way to recover moral and political confidence.[23] This point of
possible contact with hermeneutics is given support by Strauss' acknowledgement in 1949 that in our age political philosophy has a particular need to engage in a study of history, but a non-historicist one. We
need a special kind of philosophical inquiry "to keep alive the recollec-
tion, and the problem, of the foundations hidden by [the modern belief
in] progress. This philosophical inquiry is the history of philosophy or of
science." While still rejecting historicism, which asserts that the fusion of
philosophy and history will result in a progress beyond naive or pre-
historicist philosophy, Strauss reluctantly accepts some such fusion as,
"within the limits indicated, inevitable on the basis of modern philosophy, as distinguished from premodern philosophy." The limited purposes of this fusion are: "to transform inherited knowledge into genuine
knowledge by revitalizing its original discovery, and to discriminate between the genuine and the spurious elements of what claims to be inherited knowledge."[24] Such a fusion—a concept adopted from his interlocutor, but restricted to his own purpose—is necesary in order to keep (and
only to the extent that it keeps) modern philosophy truly philosophical,
and steers it back to its correct course.

But for Strauss, that course can only be securely based on the conviction that absolute truth is real and accessible in principle to human consciousness. Consequently, no reply by Gadamer will be satisfactory until

---

21 "Correspondence," 11.

22 "Correspondence," 11.

23 At the close of his second letter, Strauss makes this gesture: "I do not believe
that either of us possesses full clarity about this issue: all the more reason that we
should continue to try to learn from one another. I promise you that I shall do this."
("Correspondence," 11.) But as we will see, his suspicion of the historicist and relativist character of Gadamer's philosophy subsequently appeared to change very little.

24 Strauss, "Political Philosophy and History," 77.

he can show that philosophical hermeneutics is not self-contradictory in the way of conventional relativism, and that it at least affirms the reality of universally valid political and moral truths. For Strauss the inherently historicist character of philosophical hermeneutics remains both theoretically incorrect and potentially dangerous in practice. So while Strauss' own ambivalence about philosophical hermeneutics suggests that his general objections to historicism can be overcome, the persuasiveness of that rebuttal will depend on whether philosophical hermeneutics can offer new grounds for confidence in moral and political truth.

DIALOGICAL ONTOLOGY

Strauss' appellation of philosophical hermeneutics as "post-historicist" is more revealing than Strauss perhaps intended. It suggests that Gadamer's perspective is neither pre-historicist in the sense of unquestioning confidence in traditional claims, nor historicist in the sense of leading to a nihilistic relativism. In fact, Strauss' letters pose an unspoken question to Gadamer: 'Can you show us that philosophical hermeneutics is something other than either traditional or historicist hermeneutics? If there is no crisis, why do you encumber your philosophy with historicist trappings when it is still basically a restatement of the purpose of traditional hermeneutics? And if there is a crisis, why don't you acknowledge and address it with a corresponding hermeneutics—and acknowledge its Achilles' heel of relativism? Or can you show a new answer to that problem, if you really do have something new there?'

To rebut Strauss' objections, the meaning of historicity must be shown to be something which conventional historicism has not grasped. We might then be able to identify a hermeneutical understanding of truth and of moral universals within the conditions of human historicity, an understanding which refutes the idea that philosophical relativism necessarily leads to practical nihilism.

While Strauss in some measure senses Gadamer's orientation, it is clear that he has not fully comprehended, on its own terms, Gadamer's understanding of historicity and truth. But it is also true, as Strauss notes, that Gadamer's letter does not clearly deal with the question of relativism. However, none of Gadamer's treatments of relativism, nor his responses to Strauss can be adequately understood without a grasp of the fundamental perspective which Gadamer brings to this question. If we are to clarify these misunderstandings and confusions, we must first return to *Truth and Method* to understand Gadamer's ontology.

The hermeneutical perspective discussed in Chapter Two rests on this ontology. A merely epistemological interpretation of the hermeneutical process would open it to eventual reduction to only a methodology, an objective which some writers seek and Gadamer is very keen to prevent. In part to avoid such a reductive prospect, Gadamer's ontology makes a case, as it were, for 'why' as well as 'how' all understanding is subject to historicity. This ontological dimension provides the basic key for our objective of conceiving a hermeneutical philosophy of practice.

Gadamer's aim in the third part of *Truth and Method* is to show, through a two-fold process, how language is the "ontological horizon" of all human understanding. First, he develops a phenomenological perspective on the unity of language and world. Second, he seeks to express this perspective in terms of—and in another sense, against—the philosophical tradition of metaphysics. Our aim in this section is to interpret the substance of these discussions.

Language and World

"Language is not just one of man's possessions in the world; rather on it depends the fact that man has a *world* at all."[25] This is the phenomenological pivot of Gadamer's investigation here. In one sense language is a possession, a tool, which mediates our interaction with the world. But the peculiarly human freedom from habitat, the freedom to choose the shape and meaning of the world around us, "implies the linguistic constitution of the world. Both belong together. To rise above the pressure of what impinges on us from the world means to have language and to have 'world.'" Language does not sever us from our habitat but simultaneously permits our participation in the world and a "free, distanced orientation" toward it.[26]

Gadamer argues that the paradigmatic scene of language is one very close to us: "It must be emphasized that language has its true being only in conversation, in the exercise of understanding between people." In fact it is in such everyday communication that we see the unity of language and world:

> Coming to an understanding is not a mere action, a purposeful activity, a setting-up of signs through which I transmit my will to others. Coming to an understanding as such, rather, does not need any tools, in the proper sense

---

25 *TM*, 443 [401]. Italics his.

of the word. It is a life process in which a community of life is lived out . . . In linguistic communication, "world" is disclosed. Reaching an understanding in language places a subject matter before those communicating like a disputed object set between them. Thus the world is the common ground, trodden by none and recognized by all, uniting all who talk to one another.[27]

Thus "whoever has language 'has' the world,"[28] and conversely, "in language the world itself presents itself." In the final analysis, our linguistic experience of the world "does not imply that a world-in-itself is being objectified." There is no such absolutely independent object. If we can use the adjective 'absolute' at all, it is only to say, as Gadamer does, that "verbal experience of the world is 'absolute.' It transcends all the relative ways being is posited because it embraces all being-in-itself, in whatever relationships (relativities) [*sic*] it appears."[29]

"If we keep this in mind, we will no longer confuse the factualness . . . of language with the *objectivity of science*." For science, to objectify something is to assert a kind of power over it, to calculate it, to use it for independently determined purposes. From this perspective, natural language is but one means, one tool, by which to objectify the world. But our natural language ability is not "merely one particular form among the many other linguistic orientations [which would include, e.g., the language of science]—[rather] they themselves remain bound up with man's life orientation."[30] This 'attitude to life' implies more than just the words we use, and indicates that the scope which Gadamer imputes to 'language' is much larger. What is paradigmatic about conversation is not the use of words but that something is being communicated; every sign, sound and symbol, every gesture, pitch and tone, every text, record and narrative are also part of language in this sense.

For Gadamer this primordial unity of language and world means something which is of great significance to our project as well as his: "the verbal world in which we live *is not a barrier* that prevents knowledge of being-in-itself but fundamentally embraces everything in which our insight can be enlarged and deepened." Gadamer thus views language very differently from those who suppose that human linguisticality is

---

26 *TM*, 444 [402, 403].
27 *TM*, 446 [404].
28 *TM*, 453 [411].
29 *TM*, 450 [408].
30 *TM*, 453 [411].

fundamentally a barrier to knowledge. He suggests that the root antag-
onist here is the concept of a "'world in itself' that lies beyond all language."[31] Gadamer counters with his view of what it is that language
presents to us:

> Rather, the infinite perfectibility of the human experience of the world
> means that, whatever language we use, we never succeed in seeing anything
> but an ever more extended aspect, a "view" of the world. Those views of the
> world are not relative in the sense that one could oppose them to the "world
> in itself," as if the right view from some possible position outside the hu-
> man, linguistic world could discover it in its being-in-itself . . . In every
> worldview the existence of the world-in-itself is intended . . . The multiplic-
> ity of these worldviews of the world does not involve any relativization of
> the "world". Rather, what the world is is not different from the views in
> which it presents itself.[32]

This passage could well have been one of those which reinforced in
Strauss' mind the relativism of philosophical hermeneutics. But to see
this relativism as a kind of fragmentation of the truth which prevents us
from arriving at some absolute 'world in itself' is to miss completely
Gadamer's point. Gadamer's own warning follows almost immediately:
to oppose such 'aspects' of the world against some 'world in itself,' one
"must think either theologically—in which case the 'being-in-itself' is not
for him but only for God—or else he will think like Lucifer, like one who
wants to prove his own divinity by the fact that the whole world has to
obey him."[33] For Gadamer, such thinking reflects a futile search for an
absolute truth and proves to be an Icarian flight. Like his use of
'language,' Gadamer's use of 'world' is not simply literal; 'world' also
refers to the whole realm of being which humans can understand, and in
this sense already reveals the ontological as well as epistemological character of his perspective. The connection between this ontology, which
clearly means to assert something as universally true, and the reality of
the relativity of our 'views' of the world, is brought to light in Gadamer's
treatment of the relationship between what is finite and what is infinite.

The origins of this theme for Gadamer are in the work of Wilhelm
von Humboldt, who described language as the "infinite use of finite
means."[34] However, Gadamer draws quite different conclusions from

---

[31] *TM*, 447 [404, 405]. Italics added.

[32] *TM*, 447 [405-406].

[33] *TM*, 406.

[34] Wilhelm von Humboldt, *Über die Verschiedenheit des menschlichen Sprachbaus*
("first published 1836"), sec. 13, quoted in *TM*, 1st edition, [399] (Note 70 cites

this than can be found in the modern science of linguistics which flows from Humboldt. Focussing on the way in which language creates an infinite variety of combinations from a finite set of words and rules of syntax, linguistics perceives a dichotomy between the form and the content of language. As Joel Weinsheimer puts it in his commentary on *Truth and Method*, this dichotomy is based on treating language "as an object in itself apart from what it means." One is then led to perceive language only in instrumental terms, and to separate language as form from the world as the source of linguistic contents. Showing that Humboldt did not actually lean in this direction himself, Gadamer rejects this form-content dichotomy.[35] Instead, the infinite character of finite language is revealed in the significant fact that, as Weinsheimer puts it, "the power of language always exceeds whatever is and has been said in it."[36] This explication summarizes a claim repeatedly and variously made by Gadamer throughout his works. In *Truth and Method* this claim is most clearly stated here:

> The occasionality of human speech is not a casual imperfection of its expressive power; it is, rather, the logical expression of the living virtuality of speech that brings a totality of meaning into play, without being able to express it totally. *All human speaking is finite in such a way that there is laid up within it an infinity of meaning to be explicated and laid out*. That is why the hermeneutical phenomenon also can be illuminated only in light of the fundamental finitude of being, which is wholly verbal in character.[37]

Gadamer is indicating the simultaneity in language of finite and infinite meanings, not merely the distinction between finite forms and infinite contents.

In another work Gadamer articulates the implication of this perspective:

> The mediation of finite and infinite that is appropriate to us as finite beings lies in language—in the linguistic character of our experience of the world. It exhibits an experience that is always finite but that *nowhere encounters a bar-*

Humboldt as shown here). *TM*, 2nd edition, 440, cites an English translation of von Humboldt's work as follows: *Linguistic Variability and Intellectual Development*, trans. George C. Buck and Frithjof A. Raven (Coral Gables: University of Miami Press, 1971).

35 Joel Weinsheimer, *Gadamer's Hermeneutics: A Reading of* Truth and Method (New Haven, CT: Yale University Press, 1985), 243, 244. 90 See *TM*, 438-443 [397-402].

36 Weinsheimer, 243.

37 *TM*, 458 [416]. Italics added.

rier at which something infinite is intended that can barely be surmised and no longer spoken. Its own operation is never limited, and yet is not a progressive approximation of an intended meaning. There is rather a constant representation of this meaning in every one of its steps.[38]

Weinsheimer's commentary on Gadamer's perspective is helpful:

> Each language presents no more nor no less than a view of the world itself, and therefore a partial view. Yet each of the many language worlds implies the one common world, independent of that particular language . . . The plurality of language worlds cannot be conceived of as relativities opposed to the one absolute world because they cannot be opposed to each other . . . [Thus,] nothing is in itself necessarily unintelligible, for language—not some abstract language in general, the silent logos, or even the divine Word—but each spoken, human language is potentially infinite.[39]

Gadamer's varied assertions of human finitude point to something more ontological than would have been conveyed by speaking only of 'historical contingency' or 'relativity.' For example: "For we are guided by the hermeneutical phenomenon; and its ground, which determines everything else, is the *finitude of our historical experience*"; "language is the record of finitude"; "the event of language corresponds to the finitude of man."[40] It is in this sense that we use the term 'historicity,' and it is in this context that it shows its real ontological meaning. 'Historicity' adds to the meaning of 'finitude' here, for it more clearly connotes the social and time-defined nature of human life than does 'finitude,' and because 'historicity' does not imply, as can 'finitude,' a sense of isolation or powerlessness, which is untrue.

There is no suggestion in philosophical hermeneutics that the reality of being historical beings is a despairing one, nor that our finitude is a tragic imprisonment. Finitude can indeed be the most bitter aspect of our being to accept; but in itself it is nothing to fear or avoid, for to be finite (and this brings us back to language) is to be in the midst of what is infinite and to know we are in its midst. In this way we begin to see how, in Gadamer's ontology, what is finite and what is infinite interprenetrate in our linguisticality. This is absurd only to the mind that presumes their disjunction, and seeks to acquire the infinite so as to manipulate or discard our finitude. "No more than an infinite mind, can an infinite will

---

38   Hans-Georg Gadamer, "The Nature of Things and the Language of Things," in his *Philosophical Hermeneutics*, 80. Italics added.

39   Weinsheimer, 246-247.

40   *TM*, 457 [415]. Italics his.

surpass the experience of being that is proportionate to our finitude." Rather, "it is the medium of language alone that, related to the totality of beings, mediates the finite, historical nature of man to himself and to the world."[41]

Language and Philosophy

Gadamer seeks to conceptualize the unity of language and world through the resources of Western philosophy. This task is made more difficult since he seeks to describe this unity of word and world through the language of philosophy, which begins to show its limits in expressing what Gadamer intends. Specifically, Gadamer aims to formulate the meaning of 'dialectic' in a way which will express the unity of language and world. The result is a profound shift in the way we should view the activity and validity of theorizing.

"If we are to do justice to the subject, the hermeneutics of the human sciences—which at first appear to be of secondary and derivative concern, a modest chapter from the heritage of German idealism—lead us back into the problems of classical metaphysics."[42] The metaphysical tradition asks about the ultimate relationship between a subject and the objects before it. Gadamer means to examine hermeneutical experience in this light: "we are trying to define the idea of belonging"—between subject and object, and also between word and world—"as accurately as possible."[43]

But Gadamer intends for his approach to this belonging to be different from two other classical metaphysical perspectives, those of Greek and Hegelian thought:

> When we thus take the concept of belonging [between subject and object] which we have won from the aporias of historicism and relate it to the background of general metaphysics, we are not trying to revive the classical doctrine of the intelligibility of being or to apply it to the historical world. This would be a mere repetition of Hegel which would not hold up, either in the face of Kant and the experiential standpoint of modern science, or primarily in the face of an experience of history that is no longer guided by the knowledge of salvation. We are simply following an internal necessity of the

---

41 *TM*, 457 [415].
42 *TM*, 460 [417-418].
43 *TM*, 462 [419].

thing itself if we go beyond the idea of the object and the objectivity of understanding toward the idea that subject and object belong together.[44]

The experienced belongingness of subject and object, when put into
conceptual expression, must be done in terms of their connection or co-
ordination, not in terms of polarity or incompatibility. When "we are
thinking out the consequences of language as medium,"[45] this belongingness is evident at several points which Gadamer has already surveyed: the belongingness between interpreter and text, between tradition
and history, and among conversational partners. This belongingness
found expression in classical and medieval metaphysics to the extent that
being and truth were also seen as belonging together: this tradition
"conceives knowledge as an element of being itself and not primarily as
an activity of the subject." At this level Gadamer sides with classical
thought against the constructions of modern Cartesian thought. "In this
[classical] thinking there is no question of a self-conscious spirit without
world which would have to find its way to worldly being; both belong
originally to each other. The relationship is primary."[46]

But beyond this older insight, Gadamer cannot share the directions in
which Greek or Hegelian thinking move. His responses to these alternatives share a common argument: language, as the expression of reasoning, participates in the very belongingness we seek to describe, but this is
forgotten, in one way or another, in the attempt to secure or guarantee
our linguistic control of that belongingness. Gadamer asserts that there is
a dialectic in hermeneutical experience and he invokes the concept of
"the speculative" to distinguish this dialectic from the classical Greek
concept, and especially from Hegel's concept of dialectic as the absorp-
tion of all human reason into the Absolute Spirit. To be speculative is
here very close to the activity of hermeneutical consciousness.

> Speculative means the opposite of the dogmatism of everyday exper-
> ience . . . A thought is speculative if the relationship it asserts is not
> conceived as the quality unambiguously assigned to a subject, a property to
> a given thing, but must be thought of as a mirroring, in which the reflection
> is nothing but the pure appearance of what is reflected.[47]

---

44   *TM*, 461 [418].
45   *TM*, 461 [418].
46   *TM*, 458, 459 [416].
47   *TM*, 466 [423].

For Gadamer the speculative or mirroring nature of speech connotes the mediation of finite and infinite, and not, as asserted in other views, a reproductive or representational creation. In Weinsheimer's explanation, speculative speech

> ... always reflects more than it says. The said reflects the unsaid; the part mirrors the whole—the whole truth that is virtually present in each act of speech ... Speech is speculative in that the finite and occasional event of speech reflects virtually the infinity of the unspoken."[48]

If philosophizing is to be done hermeneutically, then it must ultimately proceed speculatively rather than propositionally. This is where Gadamer parts from Hegel.

> In fact, Hegel's dialectic ... follows the speculative spirit of language, but according to Hegel's self-understanding he is trying to take a hint from the way language playfully determines thought and to raise it by the mediation of the dialectic in the totality of known knowledge, to the self-consciousness of the concept. In this respect his dialectic remains within the dimension of statements and does not attain the dimension of the linguistic experience of the world ...
>
> Language itself, however, has something speculative about it in a quite different sense ... as the realization of meaning, as the event of speech, of mediation, of coming to an understanding. Such a realization is speculative in that the finite possibilities of the word are oriented toward the sense intended as toward the infinite.[49]

Like the dialectic of propositional philosophy, which Gadamer sees as having incorrectly triumphed in Hegel, "so also hermeneutics has the task of revealing a totality of meaning in all its relations," that is, of seeking a unified whole of meaning about each object as well as all objects. But what should protect hermeneutics from the errors of Hegel is that in another sense all understanding remains "always a relative and incomplete movement" which cannot cause an object to "appear in the light of eternity" and indicates that there will be infinite appearances of every finite object. "Every appropriation of tradition is historically different ... Each is the experience of an 'aspect' of the thing itself."[50] The acknowledgement of this seeming paradox is something which Gadamer associates with the term 'hermeneutical consciousness,' and it tells us something about all interpretation, and thus all philosophizing:

---

48 Weinsheimer, 253.
49 *TM*, 468-469 [426].
50 *TM*, 471-473 [428-430] passim.

> The paradox that is true of all traditionary material, namely of being
> one and the same and yet of being different [i.e., that each transmitted object
> both remains the same and 'appears' differently in each interpretation],
> proves all interpretation is, in fact, speculative.[51]

By implication, philosophy which is pursued in the language of proposi-
tions is not an alternative to speculative language. Rather, propositional
philosophy already occurs within the speculative character of all language.

This affirmation of the speculative character of language leads to a
different perspective on the nature of truth.

> From this viewpoint the concept of belonging is no longer regarded as
> the teleological relation of the mind to the ontological structure of what ex-
> ists, as this relation is conceived in metaphysics. Quite a different state of af-
> fairs follows . . . The fundamental thing here is that something occurs.[52]

This means that the interpreter "is not a knower seeking an object,
'discovering' by methodological means what was really meant." Of
course this occurs, and is required by "methodological discipline." But it
is an "external aspect," the outer dimension of the event. "But the actual
event is made possible only because the word that has come down to us
as tradition and to which we are to listen, really encounters us and does
so as if it addressed us and is concerned with us . . . The questioner becomes the one who is questioned."[53] This event of encounter results in
something new coming into being.

This is the hermeneutical event proper, the paradigmatic experience
of conversation or dialogue, which now appears as an ontological event
as well as an epistemological event. Gadamer thus tries to articulate the
ontological character of the subject-object belongingness in a way that
does justice to its linguistic essence, to the fact that *human speaking is finite
yet is connected to an infinity of meaning*—a connection which is not paradoxical but necessary. This belongingness of finite and infinite in which
the coordination of subject and object is played out—"this is dialectic,
conceived on the basis of the medium of language."[54]

---

51 *TM,* 473 [430]. The translators indicate that they have introduced a neologism,
"traditionary," to refer in some cases to the content of tradition ("Translators'
Preface," in *TM,* 2nd ed., xvi.).

52 *TM,* 461 [419].

53 *TM,* 461 [419].

54 *TM,* 465, [423].

The culmination of Gadamer's argument for the speculative character of understanding is an explicit statement of the ontological claim his work rests upon, which concludes with the well-known epigram of his philosophy:

> We can now see that this activity of the thing itself, the coming into language of meaning, points to a universal ontological structure, namely to the basic nature of everything toward which understanding can be directed. *Being that can be understood is language.*[55]

By way of explanation, "This means that it [i.e., being] is of such a nature that of itself it offers itself to be understood [in language]."[56] The belongingness of truth and being appears precisely in and as the historicity of language. This idea is even more clearly evident in a subsequent essay: "Precisely through our finitude, the particularity of our being, which is evident even in the variety of languages, the infinite dialogue is opened in the direction of the truth that we are."[57] Weinsheimer offers this explication of Gadamer's claim:

> If genuine language reflects not some thing but being, then it is being that makes itself understood in language. Being makes itself accessible by presenting itself in beings that can be understood, and they are its self-presentations, its language.[58]

The speculative character of language is essential to being. That is, the manner in which we understand reality is essentially related to reality itself, our own and the world's.

> The language that things have—whatever kind of things they may be—is not the logos ousias [i.e., the substance of reason], and it is not fulfilled in the self-contemplation of an infinite intellect; it is the language that our finite, historical nature apprehends when we learn to speak.
>   . . . Speculative language . . . is not only art and history but everything insofar as it can be understood. The speculative character of being that is the ground of hermeneutics has the same universality as do reason and language.[59]

---

55 *TM*, 474 [431-432]. Italics his. Gadamer's point receives a different formulation here: "In truth hermeneutic experience extends as far as does reasonable beings' openness to dialogue." ("Afterword," in *TM*, 2nd ed., 568.)

56 *TM*, 475 [432].

57 "The Universality of the Hermeneutical Problem," 16.

58 Weinsheimer, 255.

59 *TM*, 476-477 [433-434].

Gadamer here makes reason co-extensive with language; reason is not something which transcends language and to which we may appeal to bypass the historicity of language. Nor should linguistic expression be misunderstood as a separate being, into which the mirrored reflection of reality can be compressed. Of course the spoken work can be distinguished from "what comes into language in it." But the word "exists only in order to disappear into what is said."[60] Thus the speculative or mirroring character of language does not refer to something we 'put' there, but rather to something that resides in language already and therefore in being. Weinsheimer sets his own translation of two key sentences of Gadamer's within a helpful commentary on this ontology:

> "What something presents itself as, belongs indeed to its own being." (450 / 432) That this *as* belongs to its being means that (rather than being identical) it differs from itself: it is/not itself, and this is the case of everything that exists historically. It has a mirror relation to itself, a rift in its identity, which means that existing historically, as always something different, is its way of being. Yet, Gadamer writes, precisely because this is a mirror relation, the multiplicity of historical interpretations do not disintegrate into mere plurality [or, we might add, into mere 'absurd' relativity], for they are all still reflections of the thing itself. "Everything that is language has a speculative unity, a difference in itself between its being and its self-presentation, a difference however that is in fact no difference." (450 / 432)[61]

When we attempt to divorce and exclude a reality from its linguistic self-presentation in order to examine the content of the presentation, we do not gain a controlling knowledge of that reality as we sought, but instead lose awareness of our belongingness to it. Weinsheimer calls this a "difference without distinction . . . [a] unity in multiplicity . . . implied by the hermeneutic *as* which bridges (without closing) the rift in being."[62]

Gadamer's ontological claim is that the linguisticality of being is not to be feared nor to be attacked by falsely heroic effort. This is an insight which, as we will show here and in subsequent chapters, is relevant to the life of practice as well as the life of reflection. It re-presents the Socratic insight—that true knowledge is always also knowledge that we are finite—in the context of consciousness of our historicity: our finitude

---

60 *TM*, 475 [432].

61 Weinsheimer, 255. Weinsheimer's page references are to *Wahrheit und Methode* and *TM*, 1st ed., respectively; the corresponding page in *TM*, 2nd ed. (which he co-revised), is 475, where "distinction" is used instead of "difference."

62 Weinsheimer, 255, 256.

means that we are already participating in the infinite.[63] Such an ontological vision reflects not historicist anxiety nor nihilistic despair, but confidence about a universe in which we fully participate because of our finitude, and not in spite of it. This is a confidence that can afford to be humble because finitude contains no threat. It is not surprising, then, that Gadamer's philosophical hermeneutics does not follow either Heidegger's or Strauss' conclusion that we are in the midst of a world-night.

GADAMER'S REPLY

With this dialogical ontology in mind, we can turn to examine with greater understanding Gadamer's replies, both direct and indirect, to the issues which Strauss raises. We look first at Gadamer's letter to Strauss; then at certain passages in *Truth and Method* itself; and finally at Gadamer's most sustained reply to Strauss in the final pages of "Hermeneutics and Historicism."

Letter to Strauss

Gadamer's letter only indirectly addresses the general problem of relativism. His major concern is to indicate his point of divergence from Heidegger in order to correct Strauss' assumption that Gadamer must follow Heidegger in seeing our age as a world-night. But in doing so he gives a demonstration of the direction in which his ontology leads him: away from the objectivism on which Strauss depends, and towards a coordination of subject and object within history which makes sense of both human finitude and human participation in the infinite.

> My point of departure is not the *complete forgetfulness of being,* the "night of being," rather on the contrary—I say this against Heidegger as well as against Buber—the unreality of such an assertion. That holds good for our relation to tradition. We have been pressed by Schleiermacher and the ro-

---

[63] This is reinforced by the following revealing exchange between Gadamer and his interviewer, Ernest L. Fortin ("Gadamer on Strauss," 11.):

> Fortin: "You seem to regard hermeneutical philosophy as the whole of philosophy."
>
> Gadamer: "It is universal."
>
> Fortin: "Its universality implies a certain infinity; yet you insist a great deal on human finitude."
>
> Gadamer: "They go together. . ."

mantic hermeneutics into the false radicality of a "universal" understanding (as the avoidance of "misunderstanding"). I see in that a false theory for a better reality. Insofar I defend indeed the "fact of the cultural sciences"—but against itself! P. 323 intends to say in context, that *neither* the philologist *nor* the historian understand themselves correctly, because they forget the "finiteness." I do not believe in a return of pre-historicist hermeneutics, rather in its *factual* continuation, which is only hidden by history.[64]

Contrary to Strauss' perception, Gadamer identifies his perspective with pre-historicist hermeneutics, and against an historicist hermeneutics whose awareness of "history," of complete contingency, puts all claims in doubt and becomes anxiously concerned with acquiring a "universal" history. Rebuffing Strauss' implicit expectation of philosophical hermeneutics, he later adds that he does not see himself as "the prophet who already sees the promised land"; "I *remember*, instead of this, the *one* world which I alone know, and which in all decay *has lost far less* of its evidence and cohesion than it talks itself into."[65] But Gadamer's consciousness of historicity, and hence his difference from a pre-historicist hermeneutics is hinted at in the following passage of his letter:

> I have advocated against Heidegger for decades, that also his "bound" or "leap" back *behind* metaphysics is alone made possible through this itself (=historically operative consciousness!). What I believe to have understood through Heidegger (and what I can testify to from my protestant background) is, above all, that philosophy must learn to do without the idea of an infinite intellect. I have attempted to draw up a corresponding hermeneutics.[66]

The synoptic effect of Gadamer's rather diffuse letter is to suggest that the reality of historically interpretive understanding is not unique to the age of historicism, but neither should we long for the return of a pre-historicist hermeneutics as a way of avoiding the modern consciousness of history. In more substantive terms, Gadamer is opposing the idea that 'we have forgotten being,' and are thus waiting for something different, waiting for being to appear in a new way. In Gadamer's eyes, for historicism, for our age in general, and even for Strauss himself, the recovery or discovery of an "infinite intellect" is the implicit precondition of

---

64　"Correspondence," 8-9. Italics his.

65　"Correspondence," 10. Italics his.

66　"Correspondence," 10. Incidentally, this appears to be one of the few places where Gadamer speaks at all about the nature of his religious background and its influence on his philosophy.

an affirmation that being has not been forgotten. But awareness of being does not depend on this and it is arrogant to think we have that power over being—an arrogance from which even 'waiting for something different' does not necessarily escape. If we are to experience being we must put aside the idea of an "infinite intellect," with its belief in, or hope for, grasping an absolute foundation.

It is true that Gadamer does not directly address the problem of relativism which Strauss posed in his first letter. Yet in his second letter Strauss writes as though he had not even read Gadamer's careful though compressed delineation of his difference from Heidegger, which intimates a great deal of his response to that problem. Still, Strauss correctly sees in Gadamer's silence on relativism a resistance to having his philosophy reduced to either of the two categories Strauss thinks are the only possible ones: pre-historicist hermeneutics, which is (and was) occasional and for which relativism is (and was) no issue, and a hermeneutics which is radical and universal because our historicist age brings all understanding into question and now waits for some new dawn. In fact, Gadamer implicitly resists both choices because his hermeneutics is not of either sort.

Relativism in Truth and Method

There are a few passages in *Truth and Method* where Gadamer directly addresses the question of relativism. Of course, these are likely to be among the very passages which Strauss found disturbing, and wanted Gadamer to clarify. But now, with Gadamer's ontology in mind, we can begin to show why Gadamer does not regard the relativism of his philosophy, insofar as this is even an accurate label, as a problem.

Gadamer's discussion of how every assertion about the world is always a view or aspect of the world leads him to consider even the modern awareness of historical contingency as one such view, a view which characterizes our whole age. Yet, as the following statement shows, he goes on to assert the universality of this contingency.

> Even if, as people who know about history, we are fundamentally aware that all human thought about the world is historically conditioned, and thus are aware that our own thought is conditioned too, we still have not assumed an unconditional standpoint. In particular it is no objection to affirming that we are thus fundamentally conditioned to say that this affirmation

is intended to be absolutely and unconditionally true, and therefore cannot be applied to itself without contradiction.[67]

Thus Gadamer appears to accept exactly what Strauss objects to: the self-contradictory character of relativism. Gadamer goes on to comment on why he does not see this as a refutation of his philosophy:

> The consciousness of being conditioned does not supersede our conditionedness. It is one of the prejudices of reflective philosophy that it understands matters that are not at all on the same logical level as standing in propositional relationships. Thus the reflective argument is out of place here. For we are not dealing with relationships between judgments which have to be kept free from contradictions but with life relationships. Our verbal experience of the world has the capacity to embrace the most varied relationships of life.[68]

To Strauss this is precisely the kind of justification which is merely a rationalization of the absurdity of historicist relativism. But for Gadamer, Strauss' argument is precisely an illustration of how, when one aspect of the unity of language and world is taken to be the whole, we become obstinately blind to what the world-in-language really shows itself to be. Gadamer goes on to offer us a metaphor. Our experience that the sun sets beyond an unmoving horizon is not refuted by the knowledge of modern astronomy. Nor do we need to banish one or the other: just like the visual truth that the sun sets, "the truth that science states"—or, in Strauss' case, that logic states—"is itself relative to a particular world orientation and cannot at all claim to be the whole."[69]

Gadamer comments on the pointlessness of taking an aspect of the world to be the whole world:

> However clearly one demonstrates the inner contradictions of all relativist views, it is as Heidegger has said: all these victorious arguments have something of the attempt to bowl one over. However cogent they may seem, they still miss the main point. In making use of them one is proved right, and yet they do not express any superior insight of value.[70]

Such an argument, Gadamer goes on to say, has an illusory philosophical legitimacy. "In fact it tells us nothing."[71]

---

67   *TM*, 448 [406].
68   *TM*, 448 [406-407].
69   *TM*, 449 [407].
70   *TM*, 344 [308].
71   *TM*, 344 [309].

Reply to Strauss in "Hermeneutics and Historicism"

In the closing pages of the 1961 essay by this title, Gadamer responds to Strauss in a more direct and thorough fashion, both defending his views and criticizing Strauss' project. Of particular concern here is Gadamer's response to the problem of relativism's self-contradictoriness, which Strauss had pointed out in his letters.[72]

Gadamer briefly and accurately summarizes Strauss' perspective and thesis concerning the superiority of the classical vision of natural law. Moving to the issues raised in their letters, he recounts Strauss' concern over "the catastrophe of modern times": the rise of "a radical historicism that historically relativizes all unconditional values." Gadamer observes that Strauss' argument against historicism is largely based on the fact that the historicist claim itself has historical origins. He suggests, however, that this argument is itself making use of historicist thought.[73]

Gadamer goes on to directly address the problem of relativism's self-contradiction, knowing that Strauss will want to know if Gadamer is a consistent historicist or not. Gadamer obliges him: "the historical phenomenon of historicism, just as it has had its hour, could also one day come to an end. This is quite certain, not because historicism would otherwise 'contradict itself,' but because it takes itself seriously."[74] No doubt Gadamer is as familiar as Strauss with historicists who do not take

---

[72] "Hermeneutics and Historicism" was first published in 1961 in *Philosophische Rundschau*, 9 (see *TM*, 1st ed., [499], note 2, in the "Forward to the Second [German] Edition"; this information on the original publication was deleted from *TM*, 2nd ed.). It appears as an appendix to the 2nd (1965) German edition of *Wahrheit und Methode*, and to both English language editions as well (in *TM*, 1st ed., [460-491]; in *TM*, 2nd ed., 505-541.). At the conclusion of his "Forward" in the 1st English ed., Gadamer indicates that "Hermeneutics and Historicism" is reprinted "with some additions, as a supplement" ([xxvi]); this notation about the essay does not appear in the 2nd English ed. It is not known if these additions have any bearing on his discussion of Strauss at the end of this essay; in view of the way it concludes the essay, it seems unlikely that all of it was added later. And if it was indeed written in 1961, it closely follows in time upon Strauss' second letter.

In either case, the passage is primarily significant because it constitutes Gadamer's last published 'reply' to Strauss in this dialogue. Moreover, the position of the passage dealing with Strauss, as the last in a survey of literature relating to the title's topic, and the care with which Gadamer responds to Strauss' letters (without specifically mentioning them) and to the themes of Strauss' work in general (Strauss spoke in his second letter of two books of his own he was sending to Gadamer), all suggest that Gadamer thought the discussion to be of particular importance.

[73] Hans-Georg Gadamer, "Hermeneutics and Historicism," in *TM*, 533 [482, 483].

[74] "Hermeneutics and Historicism," in *TM*, 533 [483].

historicism as seriously as Gadamer does, and would, as Strauss points out, regard the disappearance of this form of historical consciousness as a terrible tragedy. (It may also be that Gadamer had in mind here the legacy of the German intellectual movement of historicism, which has a more narrow meaning than is given to the term by Strauss or by most contemporary references.) The import of Gadamer's reply is that he indicates his own particular seriousness regarding the temporality of historicism.

His seriousness is substantiated when he claims exactly what Strauss thinks is illogical, refining his position as presented in *Truth and Method*:

> Thus we cannot argue that a historicism that maintains the historical condi-
> tionedness of all knowledge "for all eternity" is basically self-contradictory.
> This kind of self-contradiction is a special problem. Here also we must ask
> whether the two propositions—"all knowledge is historically conditioned"
> and "this piece of knowledge is true unconditionally"—are on the same
> level, so that they could contradict each other. For the thesis is not that this
> proposition will always be considered true, any more than that it has always
> been so considered. Rather, historicism that takes itself seriously will allow
> for the fact that one day its thesis will no longer be considered true—i.e.,
> that people will think "unhistorically". And yet not because asserting that
> all knowledge is conditional is meaningless and "logically" contradictory.[75]

Two things stand out in this, Gadamer's most direct reply to Strauss' question. First, he does not deny that, when submitted to the same canon of logic, the two theses of historicism are contradictory. But he suggests that this case is a "special problem" because they are not "on the same level": statements referring to different orders of reflection (and perhaps of logic) cannot, he implies, be refuted because they contradict each other from the viewpoint of a single standard of logic foreign to one of them. Unfortunately he does not clearly identify these two "levels." In view of his analysis of languages as presenting aspects of the world and not the absolute world itself, it seems more accurate to understand him to be as-serting that the two theses of historicism correspond to different aspects of our experience of the world.

What he does offer, as a second point of attention, is an explanation of how, for himself as a serious historicist, the two theses are compatible and not contradictory in any sense that would require the rejection of one of the two theses. The claim that 'all knowledge is historically condi-

---

[75] *TM*, 533-534 [483]. He goes on to suggest that the end of 'thinking historically' is not just a possibility, but perhaps already upon us (534 [483-484]).

tioned' is, he says, "meaningful" even though he accepts that it was not always and may not always be considered "true." What is true is here being identified as what is meaningful; in fact, Gadamer is asserting that at least the second historicist thesis ('this piece of knowledge is true unconditionally') is meaningfully true rather than as logically true.

Yet it is just such an explanation of relativism that Strauss regards as 'hocus-pocus' to avoid the issue. From Strauss' perspective, one or the other thesis of historicism must be abandoned, the ultimate result in either case being the collapse of historicism. To Strauss, logic cannot be easily dismissed: it is rather the historicists who are not being serious and instead make a game of fuzzy experiences of "meaningfulness" in order to perpetuate the historicist illusion because they do not have the courage to acknowledge that absolute values and truths do exist.

The further unfolding of an exchange of views was, it seems, not to be. In fact, for his part, Strauss seems never to have altered his basic perception of and objection to the historicist dimension of philosophical hermeneutics. This seems clear from a passage which is worth noting since it contains what is probably his only published response to Gadamer's position in "Hermeneutics and Historicism"—and perhaps Strauss' only reference to Gadamer by name anywhere in print. Four years before his death, in a 1969 essay, Strauss implicitly discusses philosophical hermeneutics, and speaks of its unnamed author as a successor to Heidegger. And then in a footnote, Strauss refers to several pages of the second (1965) edition of *Wahrheit und Methode*, including 505, which is contained in the passage in "Hermeneutik und Historismus" in which he himself is discussed by Gadamer; thus it is clear that he had read Gadamer's reply by 1969 if not when that essay was originally published in 1961.[76] Strauss argues that for Heidegger, natural understanding is replaced by historical understanding, and that this leads to a new philosophical task, a task which is taken up in Gadamer's work: to understand "the universal structures common to all historical worlds." While this is correct, Strauss shows that he still understands "universal" in terms of his own concept of truth and thus cannot see what truth or historicity mean for Gadamer (though he reverts, at some unidentified point in the passage, to making Heidegger his object of attention):

---

[76] Leo Strauss, "Philosophy as Rigorous Science and Political Philosophy," 3. Strauss' footnote (which follows the passage we quote below) cites *Wahrheit und Methode* (2nd ed.), "233-234, 339-340, xix, and 505."

Yet if the insight into the historicity of all thought is to be preserved, the understanding of the universal or essential structure of all historical worlds
must be accompanied and in a way guided by that insight. This means that
the understanding of the essential structure of all historical worlds must be
understood as essentially belonging to a specific historical context, to a specific historical period. The character of the historicist insight must correspond to the character of the period to which it belongs. The historicist insight is the final insight in the sense that it reveals all earlier thought as radically defective in the decisive respect and that there is no possibility of another legitimate change in the future which would render obsolete or as it
were mediatise the historicist insight. As the absolute insight it must belong
to the absolute moment in history. In a word, the difficulty indicated compels Heidegger to elaborate, sketch or suggest what in the case of any other
man would be called his philosophy of history.[77]

Gadamer on His Debate with Strauss

In closing this section, it is worth noting some further information
about the correspondence between Gadamer and Strauss which
Gadamer made available in a 1981 interview conducted by Ernest
Fortin.[78] As one might expect, the question of relativism is brought up
more than once. Fortin first raises the issue to Gadamer quite un-
provocatively: "If I understand you correctly, you are reacting in your
own way against relativism. Strauss was apparently not convinced that
you had succeeded in overcoming it. Do you take his criticism to be a serious one?" Gadamer answers:

> I replied to his letter but he broke off the correspondence. I tried indirectly
> to challenge him in an appendix to the second edition of *Truth and Method*
> (pp. 482-491) [i.e., a reference to "Hermeneutics and Historicism"], but he
> did not reply to that either. We met again afterwards and I saw that he was
> very cordial.[79]

While this provides revealing personal information, at least from
Gadamer's perspective, it is striking, and even disturbing, that once
again Gadamer avoids addressing the question of relativism directly and
substantively. It can be imagined that Gadamer did not wish to resurrect
issues which opposed Strauss and himself, or which he thought would
be unfair to raise when Strauss is no longer here to defend himself. But

---

77 Leo Strauss, "Philosophy as Rigorous Science and Political Philosophy," 3.

78 Hans-Georg Gadamer, with Ernest Fortin, "Gadamer on Strauss: An Interview,"
Interpretation 12, no. 1 (January 1984): 1-13.

79 "Gadamer on Strauss," 8.

Gadamer could have avoided abusing Strauss and still grasped this opportunity to say something more substantive. As a matter of fact, his only reference in this interview (or in any published text, for that matter) to his correspondence with Strauss consists of this:

> They [i.e., the letters] revealed the strange overlapping of our positions along with a number of important divergences. The main divergence had to do with the question of the Ancients and Moderns: to what extent could this famous seventeenth century quarrel be reopened in the twentieth century and was it still possible to side with the Ancients against the Moderns?[80]

As to Fortin's question about relativism, however, nothing of substance is said here.

Be that as it may, Fortin brings Gadamer back to the issue of relativism again later in the interview, and once again, Gadamer directs his attention to his relationship with Strauss, and not on the issues raised in their philosophical dialogue. Fortin asks if Gadamer still stands by his statements in "Hermeneutics and Historicism," presumably the statements about Strauss and about relativism. Gadamer simply answers affirmatively, and then goes on to say that Strauss was "very modest" and "did not like to discuss his disagreements with me." For his own part, he adds this summation: "I have always regretted that the dialogue was not pursued. I had made a new overture and he knew that a further discussion, though perhaps not a definitive one, was possible."[81] But it was not

---

80 "Gadamer on Strauss," 3. Apparently, and curiously, Gadamer does not perceive the question of relativism to be a point of "main divergence" between himself and Strauss. Or perhaps Gadamer does not see relativism as an issue standing alone, but as subsumed by the historic 'quarrel' to which he draws attention here. Gadamer concludes his point as follows (and, incidentally, succinctly summarizes a major theme of his whole corpus), without ever addressing the question of relativism:

> I tried to convince Strauss that we could recognize the superiority
> of Plato and Aristotle without being committed to the view that
> their thought was immediately recoverable and that, even though
> we have to take seriously the challenge which they present to our
> own prejudices, we are never spared the hermeneutical effort of
> finding a bridge to them. ("Gadamer on Strauss," 3.)

81 "Gadamer on Strauss," 13. As this suggests, the interview as a whole gives the impression that Gadamer wants to speak well of Strauss, and that he is mystified as well as disappointed that Strauss did not continue their dialogue.

However, it is at least open to debate as to who "broke off the correspondence." On the one hand, it is indeed true that Strauss never responded to "Hermeneutics and Historicism," in which Gadamer represented Strauss' views fairly and gave substantive responses. On the other hand, it was also Gadamer who, in his first letter

to be; perhaps unspoken reasons, or even reasons unknown to the protagonists made this a dialogue that was never adequately resolved.

An Interpretation of the Debate: The Problem of Relativity

It is obvious that within the parameters of Strauss' focus on logic as an absolute criterion of truth it is impossible to disavow the self-contradictory character of relativism's two theses ('all knowledge is conditional' and 'the knowledge that all knowledge is conditional is itself unconditional'). We have seen how Gadamer himself acknowledges this, yet insists that Strauss' critique misses the point of what the relativity of knowledge means. Gadamer suggests that the two theses of relativism are on different "levels": from this perspective, the first thesis is a kind of 'inductive' generalization based on observations of the relativity of specific pieces of knowledge, while the second thesis is a separate philosophical statement about the first. This conceptualization of two levels appears to be an attempt by Gadamer to get around Strauss' critique, or else to express his claim in a way Strauss could understand it. Of course this does not validate relativism; it is not an argument for it, but an explanation of it.

It is more in keeping with Gadamer's ontology, however, to say that both theses of relativism are of the same speculative, as opposed to propositional, character: together they are intended to express something true about the world of human understanding, but do so through an aspect of the world—the awareness of historical contingency—which cannot correspond to the world in the form Strauss requires of truth, precisely because it is an aspect of the world. Thus the truth Gadamer intends to communicate through his relativism must itself be understood as an illustration of the concept of truth defended by his ontology. Consequently, it is impossible to validate hermeneutical relativism solely within the terms of Strauss' concept of truth. However, it is possible to make Gadamer's concept of truth more intelligible and persuasive by showing how it differs from the kind of historicism which Strauss criti-

---

to Strauss, seems hardly to have perceived (or else simply avoided) Strauss' epistemological and moral concerns about relativism, it was Strauss who apparently felt compelled to respond with a second and more blunt letter, and it was Gadamer who chose not to continue this personal correspondence, but to develop instead a public response as part of his essay on "Hermeneutics and Historicism."

cizes and with which he erroneously associates philosophical hermeneutics.

In "Hermeneutics and Historicism," Gadamer mentions a distinction Strauss makes in some works between "naive" historicism and "refined" historicism (or radical historicism, as Strauss usually puts it). The former is an awareness of the historical causation and evolution of knowledge, and the latter an additional awareness, in Gadamer's words, of the "historicity" of the "knowing subject."[82] Gadamer agrees with Strauss that the naive historicist who claims absolute knowledge of historical contingency is indeed self-contradictory; historically contingent beings cannot absolutely know their own historical contingency. In this respect, Gadamer is a radical historicist who recognizes that not only is our knowledge finite but the subject seeking knowledge is finite and cannot step beyond that finitude to survey his or her situation from an absolute vantage point. Strauss is correct in his assessment of the ultimate end of radical historicism as he understands it. When the finality of finitude is apprehended as a finality to which none of our assertions can conform, the result is despair. This is not the despair of the naive historicist who sees that there is no absolute truth which humans can acquire, but the despair of those who, having seen this, have then tried and failed in their assertions about the world to remain consistent with the insight of historicism. The end of this despair is indeed nihilism.

But philosophical hermeneutics leads in a different direction from that taken by the radical historicist. As we saw in Chapter One, Gadamer ascribes normative significance to the consciousness of historical effects; he asserts that we ought to pay attention to our finitude as we go about seeking understanding of the world and history. The idea of effective history, he goes on to say, is itself not new insofar as historians have occasionally attended to the effect of their own work on history itself, for example, in study of the history of research. But here it is still regarded as "a mere supplement to historical inquiry." As we noted in Chapter One, Gadamer makes a "new demand (addressed not to research, but to its methodological consciousness)": that we invoke this effective historical approach "every time" we seek to understand a datum of tradition, in order to see it "clearly and openly in terms of its own meaning." This, he says, is a demand that "proceeds inevitably from thinking historical consciousness through."[83]

---

82 "Hermeneutics and Historicism," in *TM*, 533 [483].
83 *TM*, 300 [267].

However, Gadamer does not add the absolute imperative which other historicists often imply: that we ought to become *completely* aware of this effectiveness in history. Gadamer says that his "demand" that we nurture a consciousness of historical effects "does not mean it can ever be absolutely fulfilled. That we should become completely aware of effective history is just as hybrid a statement as when Hegel speaks of absolute knowledge."[84] We need not perfect or complete such a consciousness because we cannot. Consequently, the despair of the radical historicist over failed attempts at consistent conformity to his or her own insight regarding finitude is pointless.

Gadamer's point is that we should both cultivate effective-historical consciousness *and* accept that we cannot do so absolutely. Thus, self-understanding means both greater awareness of our own historicity, and recognition that we cannot be aware of all of our historicity. It may seem obvious to say, as we saw Gadamer point out in Chapter One, that "the power of effective history does not depend on its being recognized."[85] But it counteracts the latent presupposition (and preoccupation) of the radical historicist that the availability of knowledge depends completely on our efforts to find and secure it; in trying to make sure that nothing they say violates the claim that we only have finite knowledge of ourselves and of the world, radical historicists ironically take upon themselves an absolute and infinite responsibility for truth and knowledge.

Gadamer's acknowledgement that, in practical as well as logical terms, historicists cannot be consistent with their own claim does not invalidate the claim that finite beings only have finite knowledge. Instead, Gadamer in effect takes the historicist's inconsistency *as a clue* to the linguistic nature of finite beings. Try as they may, historicists apparently cannot remain consistent with their claim because human assertions go beyond what finite beings can be certain they know. This feature of our linguisticality leads to a general observation: in all our claims, we assert more than just what we think we know with certainty, and assert more than we are immediately aware of asserting.[86] All assertions are interpretive in both of these senses.

---

84   *TM*, 301 [268]. A statement quoted in Chapter One bears reiteration here: The incompleteness of effective historical consciousness "is due not to a deficiency in reflection but to the essence of the historical being that we are." (302 [269])

85   *TM*, 301 [268].

86   In this form, the thesis is particularly addressed to historicists. The same thesis, stated in its opposite form can be addressed to Strauss' position: we have less certain

The features of Gadamer's perspective which have been discussed here so far do not, however, appear to increase the possibilities of acquiring knowledge or truth. The realization that the power of history's effects does not depend on its being recognized gives us no more certainty about truth than has the historicist. And if we cannot always and objectively discriminate between our certain knowledge, our assertions beyond that knowledge, and our assertions of which we are unaware, this only seems to make genuine knowledge even more elusive than it is for the radical historicist. These perspectives on the linguistic character of human assertions and on their locatedness in history seem, in fact, to demolish any possibility of truth.

The significance of Gadamer's ontology lies in its reversal of the historicist's question about knowledge and certainty; rather than asking "How can we know anything if we cannot even truly know our own finitude?" this ontology asks, "In view of our finitude, how is it that we *do* know what we know about the world?" The hermeneutic ontology is based on a phenomenological inference: our finitude is not a barrier to knowledge and truth but is the concomitant of our participation in knowledge and truth. Our understanding of the world is due to the fact that through our linguisticality we are located within a world where understanding is already happening and that we are involved in relationships with the very things about which we want to know. Our finitude is defined then not so much by our inability to ever attain absolute knowledge of our finitude as by the impossibility of relocating ourselves outside of this world and its relationships where knowledge is already occurring. Gadamer asserts that "historically effected consciousness"—though here it is better to read it as 'consciousness of historical effects'—"is so radically finite that our whole being, effected in the totality of our destiny, inevitably transcends its knowledge of itself."[87] Our being—not our finite selves but the whole of being in which we participate—is larger than the reach of a finite consciousness which aims to understand the nature of our finitude. Gadamer's point, therefore, is not that his ontology gives us more knowledge, or more certainty about our knowledge, than finite beings could otherwise have, but that it illuminates how we participate in the world, have true knowledge about it, and invoke that truth more than we are aware of at any one time.

---

knowledge than what we actually assert, and what we think we are asserting is less than all of what we are asserting.

[87]   *TM*, xxxiv [xxii].

From this perspective, truth is not something which objectively cor-responds to a separate reality, but is an event in a relationship. Truth is not what we arrive at after interpretation, but is knit into our interpreting of the world. Rather than a propositional assertion, truth is a speculative assertion, in which, as Gadamer understands it, a finite assertion pre-sents to us, through continual inquiry, the infinite meaning of a whole reality. Truth, then, may indeed be located in what we think we know objectively, but may also appear in what we are asserting beyond that knowing, and even in what we mean without self-awareness of it. For these are all ways in which finite beings find themselves to be in linguis-tic relation with others and with all of reality. Correspondingly, the world and history in which we live shape us and address truths to us for our inquiry: effective history is a way of speaking about this 'other half' of each person's dialogue with the world. Truth is not static but dynamic, and in the course of linguistic interaction is continually discovered and reforged, while pretenses to truth are exposed, inadequate assertions cor-rected and the finitude of human understanding acknowledged. The re-sults of this dialogue are more than just a neutral understanding of the world; they include the truths by which humans understand their own meaning and the meaning of reality. For Gadamer, this is the perspective that authentically acknowledges the historicity of both speakers and their claims, and in our view, has more clarity about the nature of human finitude than does either Strauss or radical historicism.

Of course, what perhaps most troubled Strauss was the fate of moral integrity in our age of anxious uncertainty. He was concerned that we in-evitably will fail to make the universal moral claims which we ought if we also affirm relativism. By contrast, the radical historicist is concerned that if we make universal moral claims we will deny the historicity and relativity of our knowledge. Now, the hermeneutical concept of truth implicitly claims to simultaneously affirm historical relativity and the possibility of true moral claims. The viability of this hermeneutical claim can be demonstrated in its response to these differing concerns.

The relativity of moral claims does not mean, as Strauss may think, that moral claims ought not to be asserted and applied universally, but rather that we should remember that such assertions are events among other events. Moral truth is indeed present in history, and therefore the relativism of hermeneutics is not nihilistic. Every assertion by finite be-ings exists in mutually influential relationships with assertions by other finite beings, both those assertions that precede it and those it will en-counter in the future. In short, effective history relativizes our universal moral claims whether we wish it or not. There is no reason such relativity should force us to relinquish our moral claims, nor does it reduce them to arbitrary or decisionist assertions. Instead our claims are fittingly understood and realized, as we saw in Chapter Two, by applying them—by concretizing them in practical judgment. The significance of Gadamer's ontology for theorizing is that it shows how every theoretical judgment is not merely available for practical concretization, but is itself the product of an earlier such concretization.

For his part, Strauss would consider such theorizing to be unreliable because no objective warrant for its true correspondence to reality has been given. However, as we have seen, for philosophical hermeneutics there are no such warrants that absolutely prove a claim to be true. But the absence of such absolute warrants does not mean the absence of genuine truth. The communicability and commonalities among societies, traditions and philosophies indicate that we live in a shared, single reality and can articulate share-able truths about that reality. The plurality of truths about this reality indicates the freedom finite beings have to understand reality in infinite possible ways. From Strauss' perspective it might also be thought that with this view of truth, philosophical hermeneutics implicitly denies that there is more to reality than what humans can understand of it. But this is not necessarily so, for as we have noted, philosophical hermeneutics is intended to be a phenomenology of human understanding, and as such does not address what is beyond human understanding. In fact, to say there is nothing beyond humanly understandable reality would itself be a claim to knowledge which humans do not have.

The radical historicist, by contrast, is concerned that by asserting moral claims we would deny and conceal the finitude and relativity of those claims. To this concern, philosophical hermeneutics replies that finitude also means that we always assert more than we are certain of, and even assert more than we are aware of. This interpretive and generative character of assertions applies to moral claims as well, and so we find ourselves already participating in moral knowledge and truth. The same 'reaching beyond ourselves' seems also to characterize the scope of applicability of our moral claims: no matter how narrowly we intend to confine the scope of moral judgment, we necessarily imply some kind of universal claim. Therefore, radical historicists cannot avoid asserting moral claims universally even if they do not intend to do so. In fact, we would be denying our finitude as much by trying to avoid making moral claims as we would by dogmatically asserting that our claim alone is the way to certain moral good. That we make moral claims, then, is not a reason for despair, on the grounds that we cannot remain 'infinitely' consistent with the historicist's knowledge that we are finite, nor for making moral judgments reluctantly, as though we must be pushed into accepting a presumptuous power to judge. Moral assertions are not the monological claims they appear to be from the objectivist perspective on truth assumed by the radical historicist as well as by Strauss. Moral assertions are always part of ever larger dialogues in which our assertions meet responses which correct and enrich them.

Gadamer's concept of hermeneutics and therefore of historical understanding can now be seen, as intimated earlier in this chapter, to be one which shares elements with both Strauss and historicism yet is also different from each. Like Strauss, Gadamer says that historicism "does not at all seem best characterized as applying the superior perspective of the present to the whole of the past," and agrees that "we must not be led into the error of thinking that the problem of hermeneutics is posed only from the viewpoint of modern historicism" but is present in every age.[88] "Under the weight of the false methodological analogies suggested by the natural sciences, 'historical' hermeneutics is separated far too much from 'pre-historical' hermeneutics."[89] But Gadamer also asserts in *Truth and Method* that "real historical thinking must take account of its own historicity,"[90] and Gadamer could be directing this at Strauss as well as at the historicist insofar as both perspectives are distorted—are slanted—by the objectivist prejudice of the Enlightenment. For Gadamer, "true historical thinking" is essentially founded on recognition of our finitude, which must especially apply in the account we give of this thinking:

---

88   "Hermeneutics and Historicism," in *TM*, 534, 537 [484, 486]. In Chapter One, we saw that in *TM*—and therefore prior to his correspondence with Strauss—Gadamer makes clear his objection to historicism: "Historicism, despite its critique of rationalism and of natural law philosophy, is based on the modern Enlightenment and unwittingly shares its prejudices." (*TM*, 270 [239].)

89   "Hermeneutics and Historicism," in *TM*, 537 [487]. Gadamer explicitly acknowledges this area of agreement with Strauss: "I do not seriously differ from Strauss, in as much as he also regards the 'fusion of history and philosophical questions' as inevitable in our thought today. I agree with him that it would be a dogmatic assertion to regard this as an absolute prerogative of the modern age." (*TM*, 536-537 [486].) However, Gadamer somewhat overstates his case here insofar as this inevitable fusion (which may be referring to Strauss' essay "Political Philosophy and History," 77) is more of a tactical concession on Strauss' part than the fundamental affirmation it is for Gadamer.

90   *TM*, 299 [267].

Historical thinking has its dignity and its value as truth in the acknowledgement that there is no such thing as 'the present', but rather constantly
changing horizons of future and past. It is by no means settled (and can
never be settled) that any particular perspective in which transmitted
thoughts present themselves is the right one. The 'historical' understanding,
whether today's or tomorrow's, has no special privilege. It is itself embraced
by the changing horizons and moved with them.[91]

Like Strauss, Gadamer asserts that the present age has no special privilege, as the naive historicist asserts. But against Strauss, Gadamer also
asserts that even in a pre-historicist age, when one's predecessors are
seen as contemporaries, "the question of truth" is still an essentially interpretive task.[92] This is a question, then, which every age must address,
not as though it starts with a blank slate, but from its own location in history.

Despite his references to classical insights about human finitude,
Strauss implicitly hopes to acquire absolute truth since he looks to the
classical age, or as we have seen possibly to a future age, which may be
especially favored with the closest approximation of this truth. And in
doing so, he reveals a measure of anxiety and self-contradiction not unlike that of historicists. In this respect it is Strauss himself, and not (as we
saw Strauss assert) Gadamer, who wants "the final insight," "the absolute insight." Precisely this conception of the purpose of hermeneutics is
what Gadamer would regard as a failure to take seriously the finitude of
our knowledge about our own finitude. Since no age is privileged in its
access to truth, there is no reason to attempt to live through an age that is
not one's own. In fact, the insights of other ages can only be interpreted
after acknowledging one's locatedness in one's own age.

Gadamer's ontology uses the language of our age of historical consciousness to try to say something about the finitude of human being
which by its nature is beyond complete understanding and is easily overlooked in this or any age. Philosophical hermeneutics presents to us an
'aspect' of the world, an aspect bearing the marks of this age. Yet this aspect also presents to us the real world. Gadamer does not claim that his
understanding of historicity is a higher truth than the understanding of
human finitude in other ages, but that it is a more thorough appraisal of
it for our age than that offered by Strauss or by historicism. A radical historicist hermeneutics is inadequate, not because it dangerously promotes

---

91 "Hermeneutics and Historicism," in *TM*, 534-535 [484].
92 "Hermeneutics and Historicism," in *TM*, 537 [486].

the illusions from which we need to be rescued, as Strauss implies, but because we need not fear our historicity, as the radical historicist ultimately does. For historicity is not a prison but the scene of dialogue where truth can and does appear despite the illusions we create.

> Does not such dialogical unendingness at its most radical signify a complete relativism? . . . In the end, though, it is also this way with the acquisition of experience: A full set of experiences, meetings, instructions, and disappointments do not conjoin in the end to mean that one knows everything, but rather that one is aware and has learned a degree of modesty . . . Taken from this perspective, hermeneutic philosophy understands itself not as an absolute position but as a way of experience. It insists that there is no higher principle than holding oneself open in conversation . . .
>
> It seems to me that the return to the primordial dialogic of the human experience of the world is irreducible.[93]

If relativism appears to open the floodgates of chaos, the implication of philosophical hermeneutics is that dialogue is always more powerful and more original (in both senses of 'original')—not because of our merit or any superhuman effort, but simply because that is the way of reality. In fact, relativism should imply a recognition of the significance of dialogue and an invitation to be disciplined by it; such discipline by dialogue is another definition of application.

'Dialogue' is the most fitting term for the focus of Gadamer's ontology. It keeps before us the historical and practical character of knowledge, whether it concerns interpretation or moral judgment, whether it concerns the science of such activities or the activities themselves. This ontology attests to a confidence about the reality of truth in history which dispels the nagging anxiety of historicism and the despair of nihilism. It is unnecessary to cling either to historicism, which seems unable to recognize the arrogance of its own attempt at humility, or to Strauss' natural law classicism, which cannot acknowledge the historicity of moral knowledge. Nor do we need to choose, as Strauss apparently believes, which to abandon, practical (and political) philosophy, or hermeneutical philosophy.

---

[93] Hans-Georg Gadamer, "On the Origins of Philosophical Hermeneutics," 188-189, in his *Philosophical Apprenticeships*, trans. Robert R. Sullivan (Boston: MIT Press, 1985).

FOUR

Habermas and Gadamer: Practice and Good

We turn now to the more public and controversial 'debate' between Jürgen Habermas and Gadamer in the late 1960's and early 1970's. Habermas' critique is like Strauss' insofar as he sees philosophical hermeneutics as finally unable to support or affirm any normative moral claims, and considers Gadamer's ontology to be primarily responsible for this inadequacy. Furthermore, both critics seek to 'save hermeneutics from itself' by adapting it to projects they find lacking in Gadamer's work. But while Strauss envisions true hermeneutics as a pre-historicist discipline which is independent of political philosophy, Habermas seeks to fashion from Gadamer's hermeneutics an instrument for mediating between a transcendental theory of knowledge and a practical commitment to social and political emancipation. Habermas' particular concern with the problem of ideology and his doubt that philosophical hermeneutics is able to justify or guide any critique of ideology make this debate an appropriate location to see how Gadamer responds to the apparent contextual relativity of his philosophy and how he conceives of its practical relevance in general.

The Habermas-Gadamer debate consists primarily of two essays by each author, beginning with Habermas' review of *Truth and Method* in 1967.[1] A host of issues were raised in these and other related essays; our

---

[1] In chronological order, these four essays in English translation are:
Jürgen Habermas, "A Review of Gadamer's *Truth and Method*" (originally 1967), in *Understanding and Social Inquiry*, ed. Fred R. Dallmayr and

discussion of them does not necessarily follow the chronological order of their publication, nor does it attempt to deal comprehensively with all of these texts. Rather, our aim is to consider the grounds for their disagreement over the universality and normative power of hermeneutics, and over the role of hermeneutics in deliberating toward the goal of good practice. Our interest is to see whether and how Gadamer's hermeneutical approach to practical reasoning can be defended against Habermas' criticisms.

The first part of this chapter identifies Habermas' central criticisms of Gadamer, the basic premises of Habermas' own position in the debate, and his counter-proposal for a more adequate formulation of the practical relevance of hermeneutics. In the second part, we present Gadamer's responses in defense of his philosophical hermeneutics and its relevance to practice, and his counter-critique of Habermas. The third part completes our consideration of the Habermas-Gadamer debate with our own interpretation of the adequacy of the hermeneutical perspective on practice in general and on the problem of ideology in particular. The last part synthesizes the results of our inquiries into Gadamer's debates with

Thomas A. McCarthy (Notre Dame, IN: University of Notre Dame Press, 1977), 335-363;

Hans-Georg Gadamer, "On the Scope and Function of Hermeneutical Reflection" (originally 1967), trans. G. B. Hess and R. E. Palmer with minor additions by Gadamer, *Continuum* 8 (1970): 77-95;

Jürgen Habermas, "The Hermeneutic Claim to Universality" (originally 1970), trans. by Joseph Bleicher in his *Contemporary Hermeneutics: Hermeneutics as Method, Philosophy and Critique* (London: Routledge & Kegan Paul, 1980), 181-211;

Hans-Georg Gadamer, "Hermeneutics and Social Science," *Cultural Hermeneutics* [now titled: *Philosophy and Social Criticism*] 2 (1975): 307-316. (Apparently this was an address, which was given in English, since no translator is identified, and it is followed by transcripts of several responses, Gadamer's reply, and an open discussion.)

Gadamer and Habermas each wrote one additional essay about the other during this period, which technically makes them part of the 'debate': Gadamer's "Replik," 283-317, and Habermas' "Zu Gadamer's *Wahrheit und Methode*," 45-56, both in Karl-Otto Apel, et al., *Hermeneutik und Ideologiekritik* (Frankfurt: Suhrkamp, 1971), a collection of essays by and about Gadamer and Habermas and their debate. However, it is open to inquiry how much these two essays add to the sources on which we will focus. Neither of them have been translated into English, and they are rarely cited, at least by English-speaking commentators on this debate, many of whom have greater facility in German than this writer; these facts suggest that their significance may be modest. For these reasons, and because this debate is not the central focus of our inquiry, these two essays have been excluded from examination.

Strauss and Habermas as they bear on the questions which were raised in the first section of Chapter Three, concerning the theoretical viability of philosophical hermeneutics as a framework for practical philosophy.

HABERMAS' CRITIQUE

Habermas claims to share with Gadamer at least one theoretical premise that Strauss rejects: that modern inquiry into the historical contingency of human knowledge justifies the conclusion that there is no epistemological starting point which is absolute in the sense of being independent of subjective consciousness. Nevertheless, he hopes to justify an alternative which could still serve as a foundation for both objective theoretical knowledge and for the practical goal of overcoming the domination of ideology in social and political life. For this reason Habermas' work is viewed as part of the 'Critical Theory' associated with the Frankfurt school of social theorists, and is also called 'ideology critique.' He anticipates that the results of his theoretical effort will have essential relevance for this practical goal: theory should help explain the cause of and correction for the forms of human practice which result in domination.

We need not even begin to consider whether Habermas' critique of Gadamer is valid to see the modern appeal of an effort to legitimate a theory with practical intent which accepts the findings of modern historicism yet attempts to ground that theory through a transcendental-like philosophical argument. And, precisely because of the proximity of this effort to Gadamer's project in regard to its historicism and general concern with practice, their differences can appear to be peripheral at first.

Habermas' critique of Gadamer is best approached by turning first to his more immediate, practical objections and then to an identification of his deeper philosophical position, which is the basis of his own theory of ideology-critique. The practical problem on which Habermas' critique focusses is the existence of what he calls "systematically distorted communication," or simply, distorted language. In such situations, "the subject does not recognize the intentions which guided his expressive activity."[2] Habermas occasionally uses psychopathological behavior as an illustration of this phenomenon, but his central interest is in its appearance in the form of a societal ideology which conceals not only what is

---

[2]　Habermas, "Hermeneutic Claim", 189.

actually meant by each speaker, but also conceals the true state of social affairs, from both speakers and hearers.

> This is the case in the pseudo-communication in which the participants cannot recognize a breakdown in their communication; only an external observer notices that they misunderstood one another. Pseudo-communication generates a system of misunderstandings that cannot be recognized as such under the appearance of a false consensus.[3]

Habermas' point can be illustrated by communication between a slave-owner and his slaves. This communication is ideologically distorted when the slave-owner and his slaves are together convinced that "we are happy and fulfilled in our positions, and our relationship is as it should be." Everyone acts and speaks as though they consent to this when in fact this is a false consensus because the slaves were not consulted about their agreement independent of the ideology of the slave-owner and his power to enforce the apparent legitimacy of his ideology. As this illustration suggests, the occurrence of ideologically distorted language is closely related to the exercise of power. Power can use language as an instrument for deceiving others into the illusion that a situation to which they did not freely agree is legitimate.

> Language is *also* a medium of domination and social power; it serves to legitimate relations of organized force. Insofar as the legitimations do not articulate the power relations whose institutionalization they make possible, insofar as these relations merely manifest themselves in the legitimations, language is *also* ideological. Here it is a question not of deceptions within a language but of deception with language as such.[4]

Where language is free of distortion caused by ideological domination, Habermas accepts Gadamer's hermeneutics as an appropriate explanation of understanding. But because the real meaning of distorted communication is opaque to those who take for granted the ideology on which that communication is based, Habermas concludes that language is unable to fully recognize or adequately expose the ideological distortions it may be carrying within itself. When the immanent tradition is taken as the scope of all possible understanding, as Habermas perceives Gadamer to be doing, then there is no objective knowledge or means by which to identify and uncover the ideological elements of that tradition.

---

[3] Habermas, "Hermeneutic Claim", 191.
[4] Habermas, "Review," 360. Italics his.

Thus ideology-critique is a necessary task but one beyond the "limits of hermeneutic understanding" as Gadamer formulates it.[5]

This is Habermas' basic objection to Gadamer. A theory of hermeneutics like Gadamer's which cannot establish some autonomy from the tradition which it interprets succumbs to a contextual relativism. It cannot be distinguished from the tradition it interprets and thereby becomes unavailable to the practical project of emancipating us from 'traditional' claims which are false but appear to be legitimate.

This limitation of philosophical hermeneutics does not, however, constitute the last word on the use of hermeneutics for Habermas. Hermeneutics must be incorporated into some larger, independent framework. The recognition of the role of power in ideology is a clue to the necessity of an extra-linguistic perspective that can understand the forces which shape language, while also understanding (as philosophical hermeneutics does) how language shapes the world for us. "Hermeneutic experience that encounters this dependency of the symbolic framework on actual conditions changes into critique of ideology."[6] And when "hermeneutical understanding . . . has been extended into critique," it no longer needs to be "tied to the radius of convictions existing within a tradition."[7]

Habermas rests his claim for the possibility of such a critical theory of ideology on the philosophical position that reason gives us a standpoint which is not confined to or defined by any particular tradition or language. This claim is, in some sense, a transcendental claim for reason. By comparison, Gadamer has excluded this power of reason from his hermeneutics, and, in Habermas' view, thereby "imposes fundamental restrictions upon the interpreter's commitment to enlightenment."[8] But Habermas also wants to affirm that reason is related to language in some way. Here he explains this relationship:

> Languages themselves possess the potential of a reason that, while expressing itself in the particularity of a specific grammar, simultaneously reflects on its limits and negates them as particular. Although always bound up in language, reason always transcends particular languages; it lives in lan-

---

5 Habermas, "Hermeneutic Claim," 189, 190.
6 Habermas, "Review," 360.
7 Habermas, "Hermeneutic Claim," 208.
8 Habermas, "Hermeneutic Claim," 208.

guage only by destroying the particularities of languages through which alone it is incarnated.[9]

Habermas calls reason an "intermittent generality" in linguistic experience (apparently to avoid classical transcendentalism), but it is "formally reflected in a characteristic that is common to all traditional languages and guarantees their transcendental unity, namely, in the fact that they are in principle intertranslatable."[10] Thus rational reflection is involved in every form of communication, contemporary or historical, between languages or within a single one, and it is this power of reason—embodied in language yet essentially beyond it—which recognizes disturbances in communication and is able in principle to restore full understanding.[11]

Critical reason penetrates and corrects distorted language through the application of Habermas' concept of truth. He conceives of truth as the social consensus derived from communication which is free of the powers which enforce ideology and distort language. This concept of truth emphasizes the formal rather than substantive significance of transcendental reason. When distorted language is subjected to rational critique, it does not necessarily tell us what is substantively true or good, for this would presuppose that the voice of reason is everywhere the same and would contradict Habermas' acknowledgement of the historical contingency of each language and practical situation. Instead, critical theory claims to have identified a universal knowledge of the conditions of unforced communication by which one can judge whether any given social agreement is a true (free) or false (ideologically distorted) consensus.

According to Habermas, critical reason not only confirms the presence of ideological distortion, but makes possible a systematic theory of language which is able to acquire knowledge beyond the competence of ordinary language and therefore beyond the scope of Gadamer's hermeneutics as well. In practical application, such knowledge about false or pseudo-agreements within tradition has power to emancipate persons from ideological consciousness.[12] Because "every consensus . . . is, in principle, suspect of having been enforced through pseudo-communication,"[13] critical theory has universal applicability. The general

---

9   Habermas, "Review," 336.
10   Habermas, "Review," 336.
11   Habermas, "Review," 338.
12   Habermas, "Hermeneutic Claim," 202-203.
13   Habermas, "Hermeneutic Claim," 205.

perspective which Habermas takes here has sometimes been described as the 'hermeneutics of suspicion.' Habermas himself often calls his own theory a "depth-hermeneutics" or "meta-hermeneutics," which he contrasts to Gadamer's "ordinary hermeneutical understanding."[14] In short, Habermas' meta-hermeneutical theory aims, first, to salvage from philosophical hermeneutics its assumptions about the historical and interpretive character of understanding; second, to subordinate them to the critical and universal power of reason; and third, to instrumentally apply those hermeneutic assumptions in order to penetrate the ideological claims of tradition.

> Understanding—no matter how controlled it may be—cannot simply leap over the interpreter's relationships to tradition. But from the fact that understanding is structurally a part of the traditions that it further develops through appropriation, it does not follow that the medium of tradition is not profoundly altered by scientific reflection . . . The methodic cultivation of prudence in the hermeneutic sciences shifts the balance between authority and reason. Gadamer fails to appreciate the power of reflection that is developed in understanding. This type of reflection is no longer blinded by the illusion of an absolute, self-grounded autonomy and does not detach itself from the soil of contingency on which it finds itself. But in grasping the genesis of the tradition from which it proceeds and on which it turns back, reflection shakes the dogmatism of life-practices.[15]

On the basis of this philosophical claim for the power of reason to critically reflect on tradition, Habermas makes other criticisms of Gadamer's philosophical hermeneutics, including its concept of prejudice. He finds Gadamer's defense of the positive role of prejudices, or prejudgments, in interpretation to be unjustified. A traditional claim which is presented as a prejudice can and should be submitted to the critical examination of reason, as outlined by critical theory, in order to determine whether it was validated or can be validated as the product of unforced agreement. Habermas contends that "a structure of preunderstanding or prejudgment that has been rendered transparent [by critical examination] can no longer function as a prejudice."[16] Contrary to Gadamer, a prejudice which has been submitted to reason cannot remain a prejudice, but must be either rejected as untrue because it was not freely agreed on, or accepted as true, as the result of authentic consensus. Therefore, no prejudice need be accepted merely because it is viewed as

---

14　Habermas, "Hermeneutic Claim," 200-205 passim.
15　Habermas, "Review," 357.
16　Habermas, "Review," 358.

authoritative by the tradition and society in which it appears. "Gadamer's prejudice for the rights of prejudice certified by tradition denies the power of reflection. The latter proves itself, however, in being able to reject the claim of tradition . . . Authority and knowledge do not converge."[17]

Habermas also criticizes Gadamer's resistance to the introduction of any methodology into philosophical hermeneutics. In Habermas' view a methodology would discipline hermeneutics, eliminating or reducing its contextual relativism and securing its claim to objective knowledge. Habermas claims to share with Gadamer an opposition to the tendency of scientific methodology to subordinate everything to its objectivistic canons of valid knowledge. But, he asserts, this correct argument gives "no dispensation from the business of methodology in general."[18] Instead of disciplining hermeneutics by binding it to a trustworthy method, Gadamer merely subordinates it to 'tradition' in a way that emasculates its power to critically reflect on tradition. Consequently, "Gadamer unwittingly obliges the positivistic devaluation of hermeneutics."[19] It would suit the advocates of methodologism to have hermeneutics relegated to abstract realms which are inaccessible to scientific method and therefore could be ignored. But Habermas does not wish to leave hermeneutics to this fate:

> This correct critique of a false objectivistic self-understanding cannot, however, lead to a suspension of the methodological distanciation of the object which distinguishes a self-reflective understanding from everyday communicative experience. The confrontation of "truth" and "method" should not have misled Gadamer to oppose hermeneutic experience abstractly to methodic knowledge as a whole.[20]

Habermas aims to present in his own theory the methodology he finds lacking in Gadamer's. The means and objective of such a methodology are "a controlled distanciation [that] can raise understanding from a pre-scientific experience to the ranks of a reflected procedure."[21] This will elevate understanding to the superior level of scientific knowledge, a level which cannot be ignored or disputed, as can the claims or applications of Gadamer's hermeneutics.

---

17   Habermas, "Review," 358.
18   Habermas, "Review," 356.
19   Habermas, "Review," 355.
20   Habermas, "Review," 355-356.
21   Habermas, "Review," 355.

Habermas asserts that the inadequacies of Gadamer's position with regard to prejudice and methodology, and in fact all the limitations of his philosophical hermeneutics, flow "not so much from hermeneutics itself but from what seems to me to be a false ontological self-understanding of it."[22] Gadamer's failure to comprehend the limitations of his own theory is ultimately due to the way he has absolutized language in his ontology. For Gadamer,

> language becomes a contingent absolute. It can no longer comprehend itself as absolute spirit; it only impresses itself on subjective spirit as absolute power. This power becomes objective in the historical transformation of horizons of possible experience. Hegel's experience of reflection shrinks to the awareness that we are delivered up to a happening in which the conditions of rationality change irrationally according to time and place, epoch and culture. Hermeneutic self-reflection embroils itself in this irrationalism, however, only when it absolutizes hermeneutic experience and fails to recognize the transcending power of reflection that is also operative in it.[23]

Habermas asserts that we are not obligated to accept Gadamer's claim to the universality of hermeneutics "if the knowing subject, who necessarily has to draw on his previously acquired linguistic competence, could assure himself explicitly of this competence in the course of a theoretical construction."[24] In fact, "the *implicit knowledge of the conditions of systematically distorted communication,*" which is "pre-supposed" by our native abilities and made explicit by meta-hermeneutics, "*is sufficient for the questioning of the ontological self-understanding of the philosophical hermeneutic* which Gadamer propounds by following Heidegger."[25] This seems to mean that the subject's native though vague and scientifically uncertain awareness of distorted language already casts doubt on Gadamer's ontological claim that ordinary language defines the full extent of human understanding. Such an ontology only ignores systematically distorted communication, when what we need is a theory that will methodologically verify the presence of distortion in concrete cases and provide the objective means by which to apply that knowledge in practice. Habermas asserts that his meta-hermeneutical theory can achieve what Gadamer's theory does not, and does so by stepping "outside" of dialogue:

> Going beyond hermeneutic consciousness, that has established itself in the course of the reflective exercise of this art, it would be the task of a philo-

---

22 Habermas, "Hermeneutic Claim," 207.
23 Habermas, "Review," 359-360.
24 Habermas, "Hermeneutic Claim," 203.

sophical hermeneutic to clarify the conditions for the possibility to, as it were, step outside the dialogical structure of everyday language and to use language in a monological way for the formal construction of theories and for the organization of purposive rational action.[26]

In tandem with his critique of Gadamer, but especially in other works since this debate, Habermas presents a number of constructive projects, all of which are in some way based on his argument for the power of reason. One project has been an examination of psychoanalytic theory as a test case of a functionally specific meta-hermeneutic.[27] Habermas has also devoted considerable attention to developing what he calls a theory of communicative competence, which aims to identify the linguistic characteristics of authentic speaking. This project represents a central objective of his meta-hermeneutic and seeks to provide rules by which to aim at free agreement and to critique the nature of distortions in ideology.[28]

Another way in which Habermas develops his critical theory is to propose a framework for the understanding of social action and of history in general, a framework which goes beyond the perceived inadequacies of Gadamer's exclusive focus on language. We have seen how Habermas argues that language is not the only reality which shapes our understanding of the world, and that we should also give attention to "social power" (or domination); to these two basic realities he adds a third, "technical mastery" (often represented as labor). Domination and labor not only shape the world, but shape language as well and therefore

---

[25] Habermas, "Hermeneutic Claim," 203. Italics his.

[26] Habermas, "Hermeneutic Claim," 188.

[27] He argues that psychoanalytic theory is able to "by-pass" the interpretive limitations of natural language where psychopathological behavior is concerned. As such it constitutes a meta-hermeneutic, and, he argues, substantiates his claim to have refuted the hermeneutic theory concerning the universal adequacy of natural language without, however, violating its correct assertion that "we cannnot transcend 'the dialogue which we are.'"(Habermas, "Hermeneutic Claim," 190.)

[28] Basic to this theory are discussions of the nature of four dimensions in every authentic act of communication. Specifically, each such act must be characterized by comprehensibility of the expression, truth as the content, sincerity toward the hearer, and appropriateness with regard to the situation. See Jürgen Habermas, "What is Universal Pragmatics" (1976), in *Communication and the Evolution of Society*, 2-3. Also see Jürgen Habermas, "On Systematically Distorted Communication," *Inquiry* 13 (1970): 205-218; Jürgen Habermas, "Towards a Theory of Communicative Competence," *Inquiry* 13 (1970): 360-375; and other passages in his *Communication and the Evolution of Society* (1976), trans. Thomas McCarthy (Boston: Beacon Press, 1979).

have independent status.[29] For Habermas this reference system finds deeper significance in a separate theory of the three-fold, hierarchical 'interests of reason': the reference point of power corresponds to reason's most important interest, a moral interest in emancipation; the reference point of language corresponds to the hermeneutical interest of reason in communicative understanding; and lastly the reference point of labor corresponds to the technical interest of reason in productive making.[30] In effect, such a three-fold reference system offers a meta-ethical basis for normative judgment, a basis which Habermas clearly thinks is lacking in Gadamer's work.

While the merits or weaknesses of these projects do not directly concern our interest here, they do indicate that he continues to pursue the integration of his philosophical claim for the power of critical reason, a social scientific approach to the causes of linguistic understanding and distortion, and a practical commitment to social and political emancipation. They demonstrate an ambitious effort to reconcile universal claims for certain truths and values with the relativity and contingency of social life under the aegis of the accessibility of transcendental reason to human consciousness. Reflecting the same spirit as exhibited in his critique of Gadamer, these projects evince a confidence that adequate knowledge and analysis of the world will lead to a mastery of practical skills and make possible the achievement of moral goals.

---

29 Habermas, "Review," 358, 361. Habermas summarizes the framework for social understanding which results from this analysis and indicates its significance as a corrective to Gadamer's hermeneutics:

> *Social actions can only be comprehended in an objective framework that is constituted conjointly by language, labor, and domination* . . . Sociology . . . requires a reference system that, on the one hand, does not suppress the symbolic mediation of social action . . . but that, on the other hand, also does not succumb to an idealism of linguisticality . . . Such a reference system can no longer leave tradition undetermined as the all-encompassing; instead it comprehends tradition as such and in its relation to other aspects of the complex of social life, thereby enabling us to designate the conditions outside of tradition under which transcendental rules of world-comprehension and of action empirically change. (Habermas, "Review," 361. Italics his.)

30 See Jürgen Habermas, *Knowledge and Human Interests* (originally 1968), trans. Jeremy J. Shapiro, Boston: Beacon Press, 1971; and further elaboration and reply to his critics in his "Introduction" (1971), *Theory and Practice*, trans. John Viertel, Boston: Beacon Press, 1973, 1-40; and in "A Postscript to *Knowledge and Human Interests*," *Philosophy of the Social Sciences* 3 (1973): 157-189.

The general significance of Habermas' critical theory for our inquiry is that it challenges the ability of any hermeneutical philosophy of practice to resolve the practical problem of ideology. To use Gadamer's language, Habermas' critique implies that, by themselves, the moral universals which we might decide to apply in practice are incapable of resulting in social good and are more likely to cause practical evil when they are ideologically distorted. Given Gadamer's claim for the universality of language, and therefore of his hermeneutical perspective, he must in effect rely on tradition to critique itself, a situation which to Habermas makes identification of the ideological elements of tradition impossible; the interpretive capability of philosophical hermeneutics is completely relative to the manifest claims of the tradition in which it is located and to the powers which enforce the 'legitimacy' of those claims. Habermas' constructive position amounts to a claim that by replacing Gadamer's ontology with his own meta-hermeneutical theory of reason, hermeneutics will find its proper role within a theory of interpretation which is universal, verifiable and applicable in a way that Gadamer's is not. Even if Gadamer does not accept this particular proposal, Habermas' critique suggests that, as it stands, Gadamer's hermeneutical perspective is incapable of promoting practical good precisely where it is most difficult and most needed: in the face of ideological domination.

GADAMER'S REPLY

When Gadamer says in his 1986 "Afterword" to *Truth and Method* that he considers Habermas' critique to have raised an "essential issue," which amounts to "the weightiest objection" that philosophical hermeneutics has received, it is a significant measure of how seriously he takes this critique. The charge, as Gadamer summarizes it in that text, is that by building his theory of understanding upon his premise of the primordiality of language, he has "thereby . . . legitimated a prejudice in favor of existing social relations." Gadamer even concedes that there is a sense in which this is correct: "All coming to understanding in language presupposes agreement," and insofar as existing agreements reflect the status quo, philosophical hermeneutics tends also to presuppose that this is where we must start any inquiry.[31] But as to whether this represents, as Habermas seems to think, a conservative or even reactionary prejudice,

---

[31] "Afterword," in *TM*, 2nd ed., 566, 567.

and as to whether hermeneutics should have the kind of emancipatory intent or method which Habermas asserts—these are additional issues, and ones on which Gadamer strongly disagrees with Habermas.

Gadamer's responses to Habermas amount to an attempt to clarify his claim for the universality of hermeneutical experience, and a specification of that claim in the realm of critical and constructive reflection on tradition and on the pathologies of tradition. Gadamer defends this claim by arguing that his perspective already affirms those parts of Habermas' critique which have merit. In addition, Gadamer makes a counter-critique of the philosophical assumptions of critical theory, which he thinks lead in a theoretically false and practically dangerous direction. He also discusses the nature of rhetoric and the practical norm of solidarity, implying their superiority to Habermas' concepts of critical theory and emancipation, respectively.

These discussions appear in two major replies to Habermas and in several subsequent essays where Habermas is an implied, if not always stated, interlocutor. Nowhere does Gadamer present a systematic argument of defense or refutation, but rather articulates his own claims by way of contrast with Habermas' misinterpretation of those claims and by contrast with Habermas' own assertions. As explicated below, his responses amount to a rhetorical defense of the idea that social communication is given a more compreheneive account in his hermeneutical perspective than in the perspective of critical theory.

Language, Reason and Tradition

The premise of Gadamer's response to Habermas is his thesis that language is the basis of all human understanding. Gadamer does not attempt to validate this assertion to Habermas' satisfaction, nor does he directly deal with Habermas' rebuttal that language can become so distorted by ideology that hermeneutical interpretation is no longer sufficient to establish understanding. Instead, in discussing several of the issues raised by Habermas, he aims to cast doubt on Habermas' conceptual presuppositions about language, reason and tradition, and on this basis challenge some of Habermas' practical expectations of the critique of social agreements.

As might be expected, Gadamer presents a concept of language quite different from Habermas':

> Language . . . is not the finally-found anonymous subject of all historical processes and action, which presents the whole of its activities as observations to our observing gaze; rather, it is by itself the game of interpretation

which we are all everyday engaged in . . . This process of interpretation
takes place whenever we "understand," and especially when we see
through prejudices or tear away the pretenses that hide reality.[32]

All possible understanding, implicitly including all possible rational understanding, is accessible to natural language, not because language is a
tool by which we (or reason) control understanding, or because language
is an object we can survey, but because language is our participation in
understanding. Gadamer summarizes his claim thus:

> The phenomenon of understanding . . . shows the universality of human
> linguisticality as a limitless medium which carries *everything* within it—not
> only the 'culture' which has been handed down to us through language, but
> absolutely everything—because everything (in the world and out of it) is in-
> cluded in the realm of 'understandings' and understandability in which we
> move.[33]

In essence, this is Gadamer's ontological claim for language as the scene
of all human understanding. While this is precisely the claim which
Habermas disputes in view of the alleged inadequacy of language to
interpret distorted communication, Gadamer's other responses are
indirect attempts to show that language is indeed adequate.

We have seen that Habermas puts great emphasis on reason, as evident in various points concerning reflection, methodology, science, and
critique. A central part of Gadamer's reply is his objection to the presupposition and expectations of these invocations of reason, based on his
own view of tradition as the scene of linguistic knowledge. This objection
is initially independent of the practical problem of ideology, and affords
Gadamer an opportunity to restate a central theme of *Truth and Method* as
it bears on Habermas' critique.

> My thesis is . . . that the thing which hermeneutics teaches us is to see
> through the dogmatism of asserting an opposition and separation between
> the ongoing, natural "tradition" and the reflective appropriation of it. For
> behind this assertion stands a dogmatic objectivism which distorts the very
> concept of hermeneutical reflection itself. In this objectivism the under-
> stander is seen . . . not in relationship to the hermeneutical situation and the
> constant operativeness of history in his own consciousness, but in such a
> way as to imply that his own understanding does not enter into the event.[34]

---

[32] "On the Scope and Function," 88.
[33] "On the Scope and Function," 83.
[34] "On the Scope and Function," 85.

Gadamer's criticism of this "dogmatic objectivism" is specifically applied to Habermas' concept of reflection. He infers that for Habermas, reflection shows its true power "only when we see through pretexts or unmask false pretensions," but that when reflection is "occupied with the supposed phantom of language," it is powerless. To Gadamer, this gives reason "a false power, [in which] the true dependencies [of knowledge on the effects of history] are misjudged on the basis of a fallacious idealism." Moreover, reason is falsely posed as an "abstract antithesis"[35] to tradition. These objections to Habermas' concept of reason lead Gadamer to make more general criticisms of Habermas' use of hermeneutics, which "stands on the premise that it shall serve the methodology of the social sciences." Gadamer understands Habermas' project as one committed to "raising understanding up out of a prescientific exercise to the rank of a self-reflecting activity by 'controlled alienation' [i.e., the concept translated as "controlled distanciation" in Habermas' essay]—that is, through 'methodological development of intelligence.'" Gadamer replies: "What kind of understanding does one achieve through 'controlled alienation'? Is it not likely to be an alienated understanding?" By contrast, "the hermeneutical experience . . . is not in itself the object of methodical alienation but is directed against alienation." Consequently, Habermas' use of hermeneutics "for the purpose of sociological method as emancipating one from tradition places it at the outset very far from the traditional purpose . . . [of hermeneutics] with all its bridge-building and recovery of the best in the past."[36]

The essential problem Gadamer sees in Habermas, and the cause of Habermas' misuse of hermeneutics, is that he "overestimates the competence of reflection and reason."[37] And in doing so, he forgets that his own theory is "itself a linguistic act of reflection."[38] In Gadamer's view, it is Habermas' erroneous commitment to a scientific concept of reason that causes him to misinterpret the hermeneutical perspective which locates understanding within tradition.

> The phrase "connection with the tradition" means [not a "preference for the conventional, to which one must then blindly subjugate oneself," but], rather, only that tradition is not merely what one knows to be and is conscious of as one's own origins . . . Changing the established forms is no less

---

35 "On the Scope and Function," 88.
36 "On the Scope and Function," 84.
37 "Hermeneutics and Social Science," 315.
38 "On the Scope and Function," 86.

a kind of connection with the tradition than defending the established forms. Tradition exists only in constant alteration.[39]

Thus the social changes with which critical theory, i.e. reason, is concerned are themselves just as much a participation in tradition as are the ideological conventions it wishes to expose and correct. If our connection with tradition is universal in this way, and thus all understanding is hermeneutical, then Habermas' supposition that there are forces in life outside the scope of hermeneutical understanding is also false.

> From the hermeneutical standpoint, rightly understood, it is absolutely absurd to regard the concrete factors of work and politics as outside the scope of hermeneutics. What about the vital issue of prejudices, with which hermeneutical reflection deals? Where do they come from? Merely out of "cultural tradition"? Surely they do, in part, but what is tradition formed from? . . . Certainly I affirm the hermeneutical fact that the world is the medium of human understanding or not understanding, but this does not lead to the conclusion that cultural tradition should be somehow absolutized and fixed; to suppose that it does imply this seems to me erroneous. The principle of hermeneutics simply means that we should try to understand everything that can be understood.[40]

Gadamer's concept of language does not necessarily disregard the impact of sociopolitical forces on language. His point is rather that their influence on language becomes meaningful only as they are taken into language and its dialogical character; by implication, in order to correct ideological prejudices we can only approach such forces through language and the tradition which it embodies.

In these characterizations of his conceptual differences with Habermas, Gadamer has already implied that some form of critique of false social prejudices is necessary and that some form of social change could result from this. Gadamer makes this explicit as he more directly approaches the specific practical issue which Habermas raises, the problem of ideological distortion. Gadamer acknowledges that there are indeed "preconditions" of social life which can "hinder" communication as well as "enable" it,[41] and agrees that authoritative tradition does take on dogmatic shape "in innumerable forms of domination." Responding di-

---

[39] Hans-Georg Gadamer, "Replik," in Karl-Otto Apel, et al., *Hermeneutic und Ideologiekritik* (Frankfurt: Suhrkamp, 1971), 307; passage translated and quoted by David Hoy, *Critical Circle*, 127.

[40] "Scope and Function," 87.

[41] "Hermeneutics and Social Science," 315.

rectly to Habermas' criticism, Gadamer counters that, therefore, "it is an inadmissable imputation to hold that I somehow meant there is no decline of authority or no emancipatory criticism of authority."[42] The implication of Gadamer's claim for the universality of language is rather that the necessity and possibility of critique is not something for which we need some non-hermeneutical source of knowledge: it is "precisely the noble task of hermeneutics . . . to make expressly conscious what separates us as well as what brings us together."[43]

Quite clearly, Gadamer acknowledges that ideology poses a genuine problem; his position on it, however, is that there is a way to address it from his hermeneutical perspective. In the sources at hand, Gadamer himself rarely uses the word 'ideology' and never refers to 'systematically distorted language,' which suggests that he finds these terms to be over-generalized and false, respectively. They misconceive the problem at hand by casting suspicion on the legitimacy of all social agreements and by presuming that false agreements are unrecognizable and uncorrectable through hermeneutical processes. Instead of using 'ideology,' Gadamer refers to the claims of tradition or the claims of authority, and redefines the issue by posing the interpretive task as one of distinguishing between true claims and false claims, the latter presumably fitting the pejorative sense of 'ideology' as Habermas uses it. Unfortunately, Gadamer does not give specific examples of what he means by authority or authoritative claims.

The practical problem which Habermas' critique raises for Gadamer is how the hermeneutical process of understanding can be regarded as capable of distinguishing the true claims of tradition from the false when that process is already suffused by tradition. In response, two arguments can be discerned in Gadamer's discussion of the relationship between the critic and the allegedly authoritative claims of tradition. Gadamer's first argument is constructive, and concerns both why and how it is possible to find truth in traditional claims:

> Authority is not *always* wrong. Yet Habermas regards it as an untenable assertion, and treason to the heritage of the Enlightenment, that the rendering transparent of the structure of prejudgments in understanding should possibly lead to an acknowledgement of authority.[44]

---

42 "On the Scope and Function," 88.
43 "Hermeneutics and Social Science," 315.
44 "On the Scope and Function," 88.

Gadamer does not mean, as Habermas does, that if and when critical investigation confirms the validity of an authoritative claim (or prejudice, in Habermas' discussion) it ceases to be an authority (which one could either blindly assume to be true or suspect to be false). Gadamer is defending the idea that there are authoritative claims in a given tradition which are *beyond* the scope of critical reason to validate or falsify. Such authorities, once regarded as true, *remain* authorities: they do not become objective knowledge certified by reason, as Gadamer reads Habermas to be expecting. Of course, this is still only an assertion, not an argument.

However, Gadamer shifts the perspective by which the issue should be explored when he goes on to question "whether one can really say that decline of authority comes about *through* reflection's emancipatory criticism, or that decline of authority is *expressed* in criticism and emancipation." He even suggests that these are perhaps not the "genuine alternatives."[45] But he chooses to set aside the resolution of this matter, simply asserting in one place that reason and authority "stand in a basically ambivalent relation . . . [which] should be explored."[46] Gadamer thinks Habermas has improperly reduced the validation or falsification of authoritative claims to a matter of the causal effect of reason on a claim regarded as an object. If critique itself is perhaps less a cause than an expression of a claim's loss of authority, this suggests that the basis of authority does not lie in something absolute which we can control and direct, as Habermas seeks to direct reason. By refusing to identify the 'real' relation between reason and authority, Gadamer appears to be exercising what he considers to be philosophical prudence about a matter which is not in our control but is rather something in which we are always participating.

How then does Gadamer think we can distinguish between legitimate and illegitimate authority? He acknowledges that both appear to be accepted or obeyed by persons and by society at large, and he agrees with Habermas insofar as "such acceptance can often express more a yielding of the powerless to the one holding power." But the 'how' question here is more than just the technical matter it seems to be for Habermas. As before, Gadamer shifts the ground as posed by Habermas when he asserts that historical study of the rise or decline of authority will show that true authority "lives not from dogmatic power but from dogmatic acceptance."[47] Here Gadamer is using "dogmatic" in a positive

---

45 "On the Scope and Function," 89. Italics his.
46 "On the Scope and Function," 89, 88.
47 "On the Scope and Function," 89.

sense, one which Habermas would not accept. Gadamer explains what he means:

> What is this dogmatic acceptance, however, if not that one concedes superiority in knowledge and insight to the authority, and for this reason believes that authority is right? . . . Authority can only rule because it is freely recognized and accepted. The obedience that belongs to true authority is neither blind nor slavish.[48]

From Habermas' perspective, one problem with this statement is that Gadamer implies that people can accept as legitimately authoritative someone else's "knowledge and insight," when one ought to be able to confront this knowledge and insight for oneself and determine whether or not to accept it as authoritative. This points to the deeper problem Habermas would see in Gadamer's statement: in either case—whether one individually accepts a claim because it is authoritative, or because the person who asserts it has authority in society—this acceptance merely begs the question of what constitutes genuine and free acceptance. How can Gadamer know such acceptance is free until he knows that this freedom is not an ideological illusion perpetrated by the very claim he might 'choose' to accept?

One clue to how Gadamer would address these problems is his reference to "insight" in the preceding statement, which is elaborated here:

> The idea that tradition, as such, should be and remain the only ground for acceptance of presuppositions (a view which Habermas ascribes to me) flies in the face of my basic thesis that authority is rooted in insight as a hermeneutical process.[49]

For Gadamer, 'insight' indicates truth, and thus he is asserting that the content of true authority is not necessarily the same as the contents of tradition, or more accurately, of the prevailing or conventional claims which societies often equate with 'tradition.' The basis of true authority is not located in a 'vicious circle' of traditional prejudices, nor in the invocation of a transcendental reason in the manner of Habermas' meta-hermeneutical theory. For truth, understood as insight, is not essentially the product of an immanent process, either historical or theoretical, but, as we saw in Chapter Three, is an event in which the whole is illuminated by an aspect. Nor is insight, as we saw in Chapter One, knowledge of "this or that particular thing," but an awareness drawn from or recog-

---

48 "On the Scope and Function," 89.
49 "On the Scope and Function," 89.

nized in the experiencing of such things: "insight into the limitations of humanity," "a religious insight" into "the barrier that separates man from the divine."[50] Gadamer is indeed too simplistic in his implication that we can safely accept as true the insight of others. But even if that were sorted out, it is superseded by this more basic problem: if true authority is rooted in an insightful event, it is not something which anyone, certainly not a person in authority, 'has' and can dispense to others. The insight of one person is not privileged but is an insight anyone can have, and "acceptance" of an authority presupposes some kind of insight as its basis. The question of where the insight first appears is finally moot in any case: the insight one may have is indeed his or her own, but yet does not occur apart from reflective dialogue with all other claims, both historical and contemporary; this is the point of Gadamer's principle of historical effects.

This reading of Gadamer leads us to the second argument he presents to Habermas: it is impossible to guarantee the validity of every claim on which our own knowledge and insights hermeneutically rely. Consequently, he opposes what he sees as the necessary implication of critical theory: if any socially authoritative claim can and ought to be suspected of being ideologically distorted, this commitment to rational critique can consistently be fulfilled only by suspecting all the preceding or supporting claims which constitute the effective history of the one claim at hand. To Gadamer, such an infinite critique is impossible for finite beings who already stand within that effective history.

> A person who comes of age need not—but he also from insight *can*—take possession of what he has obediently followed. Tradition is no proof and validation of something, in any case not where validation is demanded by reflection. But this is the point: where does reflection demand it? Everywhere? I would object to this on the grounds of the finitude of human existence and the essential particularity of reflection. The real question is whether one sees the function of reflection as bringing something to awareness in order to confront what is in fact accepted with other possibilities, so that one can *either* throw it out *or* reject the other possibilities and accept what the tradition de facto is presenting—or whether bringing something to awareness *always dissolves what one has previously accepted.*[51]

---

50 *TM*, 357 [320].
51 "On the Scope and Function," 89. Italics his.

To assert the possibility of dissolving all claims in this way is to assert that we can completely abstract ourselves from our historicity, that we can attain an 'infinite intellect.'[52]

Gadamer does not systematically apply this analysis of true and false authority to the practical problem of dealing with social claims and communication which he himself would regard as forms of domination concealed by pretensions. However, the premise of such an analysis can be inferred from the above discussion of authority: true authoritative claims can be recognized and freely accepted even though we have no higher, ultimate authority which will guarantee, in the way Habermas demands, that such claims are in fact true and our acceptance is free. From Gadamer's perspective, the problem with seeking such guarantees is not only that it is impossible to provide them; the presupposition that we need them is already an abandonment of the reliability of truth conceived as insight.

In social terms, the above hermeneutical premise means that the possibilities of removing ideological distortion and finding agreement through language are much greater than Habermas thinks. This is the conclusion reflected in Gadamer's most direct statement of his disagreement with Habermas' perspective on both the problem of ideology and its solution:

> Insofar as speech and communication are possible at all, agreement would seem to be possible as well. Naturally that does not mean that agreement can be reached on the first try. Communication always demands a continuing exchange of views and statements. In any case, *it presupposes that there are common convictions one can discover and develop into a broader agreement.* Therefore, I cannot share the claims of critical theory that one can master the impasse of our civilization by emancipatory reflection.[53]

The reference to "common convictions" suggests that Gadamer believes the universality of language implies a content that is more than a formal definition of emancipation. This implication does not contradict the relativism of Gadamer's hermeneutics since it does not mean that the common convictions will be universally identical. It does mean that language

---

[52] Gadamer's point is stated in a different way here: "It seems to me that Habermas employs too narrow a concept when he speaks of cultural tradition. Our experience of things, indeed even of everyday life, of modes of production, and yes, also of the sphere of our vital concerns, are one and all hermeneutic. *None of them is exhausted by being made an object of science.*" ("Appendix II. A Letter," 263. Italics added.)

[53] "Hermeneutics and Social Science," 315. Italics added.

joins people in a commonality that is substantive as well as formal, an inference that is further substantiated in the final two chapters. The presence of substantive commonalities is implied by Gadamer's comment on Schleiermacher, which could just as well apply to Habermas: Gadamer asks "whether the phenomenon of understanding is defined appropriately when we say that to understand is to avoid misunderstanding. Is it not, in fact, the case that every misunderstanding presupposes a 'deep common accord'?"[54] This seems to mean that misunderstanding is recognizable more by its violation or concealment of something deeper that is already shared by all participants than by the absence of a formal concurrence among them. For Gadamer, then, the resolution of social misunderstanding is always possible because this 'common accord' is never utterly lost to us, but is always testified to by our linguisticality. This does not necessarily mean we always, or immediately or fully, recognize the presence of misunderstanding; this requires the same commitment and discipline as does the conscious appearance of concord. Gadamer's implied point is that whatever the nature or source of misunderstanding, it is not due to some defect in our linguisticality or to the powerlessness of language before ideology.

Gadamer's objections to critical theory also include a warning of its potentially "ominous consequences."[55] In his view, the end result of Habermas' appeal to reason as the means to certain knowledge about true and false emancipation is that critical theory itself assumes the character of an ideology:

> Inasmuch as it seeks to penetrate the masked interests which infect public opinion, it implies its own freedom from any ideology; and that means in turn that it enthrones its own norms and ideals as self-evident and absolute.[56]

Gadamer illustrates this by referring to Habermas' argument that psychoanalytic theory functions as a meta-hermeneutic, and that critical theory can do for societal ideologies what psychoanalysis does for individual psychopathologies. Gadamer agrees that psychoanalysis is fundamentally a hermeneutical process, but more in spite of than because of its claim to objective, methodical knowledge. Consequently, the monological character which psychoanalytic theory ascribes to the analyst's role in the therapeutic dyad is not a role which Gadamer thinks we

---

54   "The Universality of the Hermeneutical Problem," 7.
55   "On the Scope and Function," 88.
56   "Hermeneutics and Social Science," 315.

would want to ascribe to some 'analyst' of society. "What happens when he [the analyst] uses the same kind of reflection in a situation in which he is not the doctor but a partner in a game? Then he will fall out of his social role!"[57] Similarly, the critical theorist, with his or her claim to being able to objectively analyze society, absents himself or herself from his or social role. In fact,

> the emancipatory power of reflection claimed by the psychoanalyst is a special rather than general function of reflection and must be given its boundaries through the societal context and consciousness, within which the analyst and also his patient are on even terms with everybody else. This is something which *hermeneutical reflection* teaches us: that social community, with all its tensions and disruptions, ever and ever again leads back to a common area of social understanding (*ein soziales Einverständnis*) through which it exists.[58]

In Gadamer's view, Habermas distrusts the existence of such a common understanding, and this is the fundamental flaw in critical theory. This ultimately points toward political consequences of which Gadamer is implicitly very critical:

> The unavoidable consequence to which all this leads is that the basically emancipatory consciousness must have in mind the dissolution of all authority, all obedience. This means that unconsciously the ultimate guiding image of emancipatory reflection in the social sciences must be an anarchistic utopia.[59]

The dissolution of all authority and the achievement of an anarchistic utopia are not necessarily the intended practical goals of critical theory, but in the absence of any countervailing theoretical principle, Gadamer argues, critical theory does indeed have this ultimate effect. If there are no restraints as to which items of tradition will be subjected to 'emancipatory reflection,' or as to when they are subjected, the end result must be a society composed of atomized agents of reflection, each at liberty to challenge or disregard any or all prior points of agreement.

Solidarity and Rhetoric

It is clear that Gadamer is speaking on behalf of a reasoning about practice that is fundamentally different from what he perceives as

---

57 "On the Scope and Function," 94.
58 "On the Scope and Function," 94-95. Italics his.
59 "On the Scope and Function," 95.

Habermas' approach.[60] This invites further inquiry into what, indeed, practice means for Gadamer, and how he prefers to conceive of reflection on practice.

Gadamer does offer further ideas on these matters but in a somewhat indirect way: in his responses to Habermas and in a few subsequent essays, Gadamer gives attention to the concept of solidarity and especially to the concept of rhetoric. In most of these instances, he does not explicitly relate either concept to Habermas' work. But the contrast of solidarity to emancipation and critical theory of rhetoric to suggest that at least part of Gadamer's intent is to develop the constructive implications of his position vis-a-vis Habermas through these concepts.

In Gadamer's two major replies to Habermas, the concept of solidarity appears only twice in passing,[61] but it is given more attention in a few subsequent essays, notably in his 1976 essay, "What is Practice? The Conditions for Social Reason." Despite its infrequent appearance in Gadamer's work as a whole, it recurs more often than any other clearly normative term. Perhaps his most direct and normative use of the term appears here: "The point is that genuine solidarity, authentic community, should be realized."[62] The context in this instance is Gadamer's interpretation of Plato's practical intent hidden in the guise of his utopian *Republic*. But it is also clear that Gadamer himself is affirming solidarity or community as a goal of practice. And in view of its usage elsewhere, as we will see in the next chapter as well as here, it apparently defines for him the primary goal of social practice.

The substantive meaning of solidarity for Gadamer has its primary roots in Aristotle's ethics. Indeed, it has connections with a number of moral and intellectual virtues discussed by Aristotle. In order to better understand the connection between solidarity and the intellectual virtues in particular, Ronald Beiner's discussion in *Political Judgment* of the prac-

---

60 See, for example, the following blunt statement in the "Afterword" to *TM*, 2nd ed., 555-556:

> Apel, Habermas, and the representatives of 'critical rationality' are in my opinion equally blind. They all mistake the reflective claim of my analyses and thereby also the meaning of application which, as I have tried to show, is essential to the structure of all understanding. They are so caught up in the methodologism of theory of science that all they can think about is rules and their application. They fail to recognize that reflection about practice is not methodology.

61 "Hermeneutics and Social Science," 311.

62 *RAS*, 80.

tical significance of these virtues is very useful. One of these virtues, *sunesis*, or understanding, is a term which in Greek usually has no moral connotation and simply means "the comprehension of the utterances of another." But, as Beiner points out, when Aristotle discusses *sunesis* as an intellectual virtue in the *Nicomachean Ethics*, it appears with and is related to two other such virtues, *gnome* (which Beiner translates as "insight, judgment, or good sense") and *syngnome* ("forgiveness, pardon or sympathetic understanding").[63] Furthermore, when Aristotle investigates the relationship between *phronesis* and each of these three virtues, "there is a difference between the two terms and yet there is something fundamental that they hold in common."[64]

These bonds are reflected in the etymology of these terms. *Gnome*, or judgment, connotes "the ability to perceive and apprehend rightly the ultimate particulars" or concrete circumstances, which require our active response. The related term, *syngnome*, "literally means 'judgment-with' or 'judgment on the side of' another person." Significantly, "this suggests that to judge [*gnome*] is to understand [*sunesis*], to understand is to sympathize" and be able to forgive (*syngnome*).[65] Beiner points out that judgment, understanding and sympathy therefore are interrelated in a way that is not apparent in English. The English translations often used for *syngnome*, eg., 'empathy' and 'sympathy,' are passive terms, related to *pathos*, or feelings which we receive or suffer, and therefore would not fall within the realm of *praxis*. But *syngnome* comes from an active root, *gnome*, or 'judgment.' Thus, *syngnome*

> implies judging, and therefore acting, for to judge-with arises out of a context of acting with. The English in failing to convey this implicit association, conveying in fact the very opposite association, likewise fails to capture any active, *praxis*-oriented dimension to sympathy.[66]

The connections between understanding, judging and judging-with become clearer now in Gadamer's commentary on Aristotle in *Truth and Method*. All of these terms, Gadamer says, have to do with "'being understanding'" of one person by another. They represent a "modification" of

---

63   The latter virtue is transliterated as *suggnome* by Beiner, but we will instead and consistently use *syngnome*, which is how it appears in the second edition of *TM* (see 323, note 264).

64   Ronald Beiner, *Political Judgment* (Chicago: University of Chicago Press, 1983), 76, 75, 74.

65   Beiner, *Political Judgment*, 76, 75. Bracketed terms added.

66   Beiner, *Political Judgment*, 76.

phronesis since "in this case it is not I who must act." But, significantly, they "stand beside" *phronesis*. Accordingly, *sunesis*[67] "means simply the capacity for moral judgment. Someone's sympathetic understanding is praised, of course, when in order to judge he transposes himself fully into the concrete situation of the person who has to act." Gadamer refers only briefly to *gnome* and *syngnome*, calling them "other varieties of moral reflection listed by Aristotle, namely insight and fellow feeling," respectively; these, he says, are illustrated in the person who is ready to forbear and forgive.[68] The point, at any rate, is that having genuine understanding (*sunesis*) in moral life is inseparably related to judgment or insight (*gnome*) and to a readiness to judge-with others (*syngnome*). Thus understanding is not mere comprehension but, as Gadamer puts it in another text, is "a kind of communality in virtue of which reciprocal taking of counsel, the giving and taking of advice, is at all meaningful in the first place."[69]

Solidarity can be regarded as a norm which invokes all three of these virtues. Furthermore, they should lead us to think of solidarity not only, nor even primarily, as a state or an anticipated goal, but as an event in which we participate. Therefore solidarity is something which requires us to decide *how* to concretely understand, judge and judge-with others, and it is the function of *phronesis* to make this decision. This makes it clear that knowledge of solidarity and the capacity to enter into it is not a quantifiable object which can be technically applied; what Gadamer says of moral knowledge applies to solidarity as well: "The question here, then, is not about knowledge in general but of its concretion at a particular moment."[70]

From this perspective, solidarity bears a close relationship to friendship, and in fact friendship should be viewed as a part of the constellation of virtues and normative universals to which Gadamer appears to be pointing through the use of the term 'solidarity.' In at least one place Gadamer explicitly affirms and joins the concepts of solidarity and friendship, and by implication judging-with as well, as the basis of societal networks of community.

---

[67] *TM*, 2nd ed., 323, uses *"synesis"* once, though the distinction, if any, is not made clear.

[68] *TM*, 322-323, note 264 [288, note 228]. Gadamer indicates in the footnote in the 2nd edition that he has "slightly revised the text here." While it is true that *gnome* and *syngnome* are somewhat clarified, the treatment remains quite brief.

[69] *RAS*, 133. See also a parallel at "Practical Philosophy as a Model," 84.

[70] *TM*, 323 [288].

> When I say that just friends are able to give good advice, I take this to mean that the concept of friendship should be expanded as much as the concept of communicative understanding. And what does a society mean without patterns of self-evident solidarity between human beings, neighbors, members of a family, colleagues in a profession—in every case, a common basis of solidarity?[71]

His statement about "communicative understanding" seems to refer to Habermas' theory of communicative competence, and implies that without the virtue of friendship, such a theory lacks recognition of an essential element of practice, an element for which no theory can substitute.

Unfortunately, Gadamer gives little further attention to friendship, especially in terms of Aristotle's treatment of it as a moral virtue, and it deserves more in view of its relevance to the development of the hermeneutical philosophy of practice. Again Beiner is of aid here, as he discusses the connections between political friendship and political judgment. He reminds us that the meaning of friendship for Aristotle "is a matter of community, and community, in turn, is a matter of justice. Thus friendship is defined as sharing a common view of what is just." The fruit of such political friendship is represented by Aristotle's term *homonoia*, or concord, literally "'being of the same mind', 'thinking in harmony,'" not in all things but in political things.[72] Beiner's conclusion about friendship adds to Gadamer's analysis and can help guide our subsequent reflection: "the quest for a theory of political judgment leads irresistibly to the formulation of a corresponding theory of friendship. To judge is to judge-with, [and] to judge-with is to be a friend."[73] By incorporating these thoughts into the concept of solidarity, we can see that it provides a loose moral definition of good society as a community of friends who share and apply a common concept of justice by virtue of judging-with each other.

Gadamer has not given systematic attention to interpreting the relevance of Aristotle's ethic for the practical issues posed by Habermas' critique. Nor in terms of comparison has he done more than imply that critical theory lacks attention to the role of friendship in social practice. Still, it is significant that by his use of solidarity, Gadamer is clearly affirming a moral universal of some kind, a universal which we ought to apply in practice, and regards it as a proper part of his philosophy as it bears on practice. His affirmation of solidarity in this way is perhaps partly in re-

---

71 "Practical Philosophy as a Model," 84.

72 Beiner, *Political Judgment*, 79, 80.

73 Beiner, *Political Judgment*, 82.

sponse to those who either were concerned about the absence of such a commitment, or doubted whether is could be made in the context of a philosophy of hermeneutics. Hereafter we will use 'solidarity' to refer to the constellation of virtues and norms which have been discussed here.

One further statement by Gadamer about solidarity is worth attention because it concerns not so much its thematic content but its role in practice. It appears at the end of "What is Practice?" where he answers the essay's title question in a seemingly enigmatic fashion:

> Practice is conducting oneself and acting in solidarity. Solidarity, however, is the decisive condition and basis of all social reason.[74]

The first sentence confirms that solidarity constitutes Gadamer's summary definition of good practice, and of the good at which practice aims: solidarity is the goal or 'effect' of good practice. But the second sentence implies that good practice depends on solidarity having been already achieved, since he says solidarity is the "basis" of social reason, which is the scene of practice. Taken together, then, these statements suggest that solidarity is both the condition of practice and its goal. This makes no sense if we presume that one cannot already have that at which one aims, but must causally realize the goal through other means. But this is the sort of technical thinking about practice which Gadamer seems intent on opposing here: social reason is not, in the final analysis, the precondition and basis of solidarity; rather, solidarity is the "decisive" basis of social reason, and of the freedom and agreement Habermas seeks. From this perspective, genuine solidarity is not something produced by formal conditions, nor through the power of a theory; it is neither the form nor the content of a thing. It is something which occurs when, through insight and judging-with, we find a common reality with others, and is confirmed when it can be freely shared. Understanding of and action toward the goal of solidarity primarily depends on practicing solidarity.

To understand 'judging-with' you must practice 'judging-with.' Since Gadamer seems to suggest that it is possible to presuppose in some way the very thing at which practice aims, the implication is that we are already unaware participants in some kind of solidarity over which we do not have control, but on which we can draw and to which we can add in our conscious actions. This reading of Gadamer brings out more than the point that practice is not a technical act and that theorizing is not the basis of its success. It identifies some kind of bond between his description

---

[74] *RAS,* 87.

of the ontological basis of human understanding and his affirmation of solidarity as a moral universal. The possibility of solidarity, Gadamer appears to mean, is 'built into' our linguistic nature: to communicate is to do more than comprehend, it is to understand and have insight into the reality of others; this insight makes it possible to 'judge-with' others, and this is the beginning of solidarity and community.

It is this bond between our linguisticality and the possibility of solidarity which is the latent basis of Gadamer's discussions of a second topic, the study and activity of rhetoric. As in his references to solidarity, Gadamer's discussions of rhetoric are not directly or systematically compared with Habermas' project. But the contrast is quite evident, and since part of these discussions occur in his more direct responses to Habermas, Gadamer's intention to point to the differences between rhetoric and Habermas' critical theory is very clear.

But why would Gadamer leave hermeneutics to discuss rhetoric? At first sight, rhetoric and hermeneutics appear to be quite different disciplines and activities. Hermeneutics is originally an historical and textual discipline, concerned with the problem that "something distant has to be brought close, a certain strangeness overcome, a bridge built between the once and the now." By contrast, rhetoric is an essentially practical activity, aimed at learning to use speech to effect persuasion in one's partner or audience. In a rhetorical situation, "the convincing power of . . . [the orator's] arguments overwhelms the listener.[75] Gadamer's interest in rhetoric lies in its capacity to permit speaker and hearer to discover their communality and common convictions in a way that includes but exceeds critical examination; this proves to be the basis of its affinity with hermeneutics.

Gadamer invokes the work of both Plato and Aristotle to defend this view of rhetoric. He describes Plato's *Phaedrus* as "dedicated to the task of endowing rhetoric with a more profound meaning and of allowing it a share of philosophical justification." Plato concluded that when claims are made for rhetoric that rival the universality of philosophy, as they were in Plato's time, rhetoric must be made "to overcome the narrowness of merely rule-governed teachings," and thus "it ultimately has to be taken up into philosophy, into the totality of dialectical knowledge." In this sense "rhetoric shares with dialectic the universality of its claim." Gadamer also refers to Aristotle's *Rhetoric* as a work "which presents

---

[75] "On the Scope and Function," 81. Gadamer adds that the listener "cannot and ought not to indulge in a critical examination," although we have seen that Gadamer does not intend to discourage all critical judgment.

more a philosophy of human life as determined by speech than a technical doctrine about the art of speaking."[76] These classical views indicate to Gadamer that when it is an authentic discipline, rhetoric is *simultaneously* rooted in both the dialectical knowledge of ultimate universals and the practical knowledge of good in ultimate particulars. From this classical perspective, in other words, "rhetoric is indissoluble from dialectics; persuasion that is really convincing is indissoluble from knowledge of the truth.[77]

Such a concept of rhetoric, Gadamer asserts, has a "very close relationship" with hermeneutics which builds on the fact that "the ability to speak has the same breadth and universality as the ability to understand and interpret."[78] More pointedly, hermeneutics and rhetoric share "the realm of arguments that are convincing (which is not the same as logically compelling)."[79] It is worth noting that the passivity of interpretation which Gadamer often emphasizes in *Truth and Method* is counterbalanced in sources such as these when he highlights the commonality between hermeneutics and rhetoric, implying that interpretation is a kind of productive oration on behalf of the text's truth.

Our concern in this chapter with Gadamer's perspective on practice focusses attention on certain additional features of the commonality between rhetoric and hermeneutics which Gadamer stresses. These features concern the fact, noted above, that for Plato and Aristotle, rhetoric participates in the highest aim of knowledge, the knowledge of good. But it is the *kind* of knowledge and participation in the good that is of particular importance. Gadamer emphasizes that, like hermeneutics, the knowledge and discipline of rhetoric is something which depends in the first place not on *techne*, but on practice. To take oratorical skill for example,

---

[76] *RAS*, 118, 119, 120.

[77] *RAS*, 122.

[78] *RAS*, 119. In the interpretation of texts for example, "the grasping of the meaning of the text takes on something of the character of an independent productive act, one which resembles more the art of the orator than the process of mere listening. Thus it is easy to understand why the theoretical tools of the art of interpretation (hermeneutics) have been to a large extent borrowed from rhetoric." ("On the Scope and Function," 82.) In broader terms, if the discipline of rhetoric is more than technique and includes "true knowledge," then the same "must be allowed to apply to hermeneutics as the art of understanding." (*RAS*, 120.)

[79] "Afterword," in *TM*, 2nd ed., 568. He goes on to say that this "is the realm of practice and humanity in general . . . [not] where emancipatory reflection is certain of its 'contrafactual agreements,' but rather where controversial issues are decided by reasonable consideration."

Gadamer observes that "if natural giftedness for speaking is lacking it can hardly be made up for by methodological doctrine."[80] This directs us to a more general point:

> What is at issue for the truly dialectical rhetorician as well as for the states-man and in the leading of one's own life is the good. And this does not pre-sent itself as the *ergon* [i.e., product], which is produced by making, but rather as *praxis* and *eupraxis* [i.e., good practice] (and that means as *en-ergeia*).[81]

Rhetoric lies in practice because it participates in the 'energy' which lets good occur, and not the 'energy' that makes a product (*ergon*). Rhetoric is a channeled 'explosion' of energy by which an audience discovers its communality, not a channeled 'implosion' of energy focussed on 'producing' an audience that thinks and feels as one, in the manner of an ideology. Gadamer elaborates on the subordinate role of theoretical rules to the essentially practical character of the discipline of rhetoric (and of hermeneutics):

> For it is clear that rhetoric is not mere *theory* of forms of speech and persua-sion; it can develop out of a native talent for *practical* mastery, without any theoretical reflection about ways and means. Likewise, the art of under-standing, whatever its ways and means may be, is not dependent on an ex-plicit awareness of the rules that guide and govern it. It builds, as does rhetoric, as a natural power which everyone possesses, more or less. It is a skill in which one gifted person may surpass all others, and theory can at best only tell us why. In both rhetoric and hermeneutics, then, theory is sub-sequent to that out of which it is abstracted—i.e., to *praxis*."[82]

The practical rather than technical character of rhetoric points to an important conclusion about the nature of good in practical dialogue:

> This knowledge of the good and this capability in the art of speaking does not mean a universal knowledge of 'the good'; rather it means a knowledge of that to which one has to persuade people here and now and how one is to go about doing this and in respect to whom one is to do it.[83]

In short, "real knowledge has to recognize the *kairos*." Once again we en-counter Gadamer's theme of practical concretization, but now in the con-text of the act of speaking. Here as well as in all practical action, "there

---

80 *RAS*, 114.
81 *RAS*, 123.
82 "On the Scope and Function," 80.
83 *RAS*, 121.

are no rules governing the reasonable use of rules":[84] there is no theoretical alternative which is superior to the practical knowledge derived through rhetorical speaking which can tell us what is the right thing to say to a given audience at a given time. Habermas seems to agree with Gadamer's point about the relevance of rhetoric throughout the realm of everyday social and practical life,[85] but he is not convinced that this realm is truly universal. Habermas argues that rhetorical (as well as hermeneutical) processes are inadequate for the purposes of bringing the monological knowledge of scientific reason to bear on everyday life.

> Hermeneutic consciousness does, after all, emerge from a reflection upon our movement *within* natural language, whereas the interpretation of science on behalf of the life-world has to achieve a mediation *between* natural langauge *and* monological language systems. This process of translation transcends the limitations of rhetorical-hermeneutical art which has only been dealing with cultural products that were handed down and which are constituted by everyday langauge.[86]

Gadamer's objection to the first sentence would be based on his sense that the second sentence completely misunderstands the scope of his philosophy when Habermas defines it as "only" concerned with the transmission of "cultural products." From Gadamer's perspective, we do not need—and cannot find—some third element to mediate between science and the rhetorical realm of social life; rhetoric *is* what interprets science and mediates it to society. "All science which would wish to be of practical usefulness at all is dependent on it [rhetoric]."[87] In an essay appearing after Habermas' statement above, Gadamer restates in more general terms his disagreement with Habermas on rhetoric:

> Rhetoric is not restricted to special institutions of our technical civilization, to the political assembly or to its technical promulgation by the mass media. Rather, it is an indispensable ferment of daily life and of the forms of communication in general. Hence, the field of rhetoric is broader than that of the sciences and technology.[88]

---

84   *RAS*, 121.
85   Habermas, "Hermeneutic Claim," 183-185, 187.
86   Habermas, "Hermeneutic Claim," 188.
87   "On the Scope and Function," 83.
88   "Hermeneutics and Social Science," 315. Gadamer elaborates on this in his first response to Habermas:

> Rhetoric . . . from oldest tradition has been the only advocate of a claim to truth which defends the probable, the *eikos* (*verisimile*), and

But Habermas' response to this view would be to ask how we can ever know that claims made through rhetoric are reliable or that the persuasion effected by rhetoric is free. Gadamer's frank counter-critique follows:

> I find it frighteningly unreal when people like Habermas ascribe to rhetoric a compulsory quality that one must reject in favor of unconstrained, rational dialogue. This is to underestimate not only the danger of the glib manipulation and incapacitation of reason but also the possibility of coming to an understanding through persuasion, on which social life depends.[89]

AN INTERPRETATION OF THE DEBATE:
THE PROBLEM OF IDEOLOGY

The question raised by the debate between Habermas and Gadamer is whether the relativity of language to its social and traditional context precludes the normative relevance of hermeneutics to concrete social practice. To know when a claim is an ideological one (as Habermas puts it) or a false authority (as Gadamer puts it) and to be able to dissolve and replace it is a central problem for practice in general and for the practice of speaking and communicating in particular. It is a fitting issue on which to test the practical relevance of the hermeneutical perspective.

Although Gadamer does not fully address the problem of ideology nor attempt to refute Habermas on Habermas' own terms, the basis of his response to Habermas in this debate is fundamentally persuasive. The dialogical ontology, on which Gadamer's responses ultimately rest, allows a more accurate depiction of the relationship between language and

---

Rhetoric . . . from oldest tradition has been the only advocate of a claim to truth which defends the probable, the *eikos* (*verisimile*), and that which is convincing to the ordinary reason, against the claim of science to accept as true only what can be demonstrated and tested! Convincing and persuading, without being able to prove— these are obviously as much the aim and measure of understanding and interpretation as they are the aim and measure of the art of oration and persuasion. And this whole wide realm of convincing "persuasions" and generally reigning views has not been gradually narrowed by the progress of science, however great this has been; rather, this realm extends to take in every new product of scientific endeavor, claiming it for itself and bringing it within its scope. ("On the Scope and Function," 82.)

[89] "Afterword," in *TM*, 2nd ed., 568.

ideology than does critical theory because it more fully understands both the historicity of human practice and the nature of the good at which practice aims. Gadamer applies his ontology in his response to Habermas through two arguments: first, that every authentic critique of ideology is doing no more than demonstrating the hermeneutic character of language; and second, that a critique which does not recognize its own hermeneutical character—something which Gadamer suggests applies to Habermas' critical theory—is ultimately only promoting a new ideology. The persuasiveness of these arguments can be substantiated by the following interpretation of both authors and of the issues they address.

It is precisely Gadamer's ontology which Habermas alleges to be the source of the powerlessness of philosophical hermeneutics when it encounters ideology. In view of the difference between their respective basic concepts, the ontological status Gadamer gives to language cannot be abstractly validated to Habermas' satisfaction. However, we can make the application of this ontology to the problem of ideology more intelligible within the terms of Habermas' position, and to the extent that this application is persuasive, bring into question the adequacy of Habermas' perspective. This is the first objective of this section. It will be followed by a discussion of the comparative merit of Gadamer's concepts of solidarity and rhetoric over Habermas' concepts of emancipation and critical theory, respectively. In the third part we consider and reject the possibility that Habermas' theoretical project is compatible with philosophical hermeneutics, even when the practical intent of emancipation from domination is affirmed as a valid norm.

The Practical Problem of Ideology

From Habermas' perspective, philosophical hermeneutics must address one or both of Habermas' central claims about the practical problem of ideology: first, that ideology is so powerful and deceptive and, second, that language is so limited and powerless, that ordinary hermeneutical processes are insufficient to recognize, penetrate and correct the misunderstanding that ensues among ideologues (who may be victims and/or perpetrators) or between ideologues and non-ideologues. If philosophical hermeneutics can convincingly oppose one or both of these claims, thereby validating Gadamer's claim for the efficacy of language in ideological situations, it would be unnecessary for Habermas to seek or to invoke a transcendental theory or source of knowledge to do what he contends language cannot do. Whether Habermas thinks there is such a theory or such knowledge would become irrelevant, insofar as the practical need it was to serve would be shown to be non-existent.

The first step toward this end is a closer examination of the implications of Habermas' distinction between ordinary (undistorted) language and ideologically distorted language. Habermas' claims amount to assertions that ordinary language is powerless to decipher the distortions of an ideologue's speech, and that the ideologue is similarly powerless to decipher the distortions in his or her own language in the course of social communication. This implies one of Habermas' premises: that the linguistic ability of all persons is virtually limited to understanding only what others say, with little or no recognition of what was actually meant if that happens to be sufficiently different from what was said.

What Habermas refers to as *systematically* distorted language seems to be a case of—or even the definition of—such a difference between what is said and what is meant. Not only is the ideologue unable to recognize the difference between what is said and what is meant (at least on matters about which ideology has distorted his or her consciousness); there is also no linguistically communicable moral or truth claim to which the critic could turn that would help reveal that difference to the ideologue. The ideologue's consciousness will distort the meaning of such claims before the critic can challenge the validity of the ideology. This implies a second premise of Habermas': that distorted language is of an essentially different order from ordinary language, and for this reason the hermeneutical process of understanding which characterizes ordinary language is not sufficient for penetrating distorted language. When what is *meant* is sufficiently different from what is *said*, the thing meant ceases to be linguistic or accessible to language.

The joint effect of these two premises is that the ordinary language speaker or would-be critic—who may, with some kind of "implicit knowledge," as Habermas asserts, at best sense the presence of distortion in the ideologue's speech but has no objective assurance of it or means to correct it—is in a position of *contextual disadvantage*. According to Habermas, no hermeneutical process will give the critic the means to awaken the ideologue to his or her distorted language. Consequently, the critic of ideology requires knowledge which is extra-linguistic or extra-hermeneutical but which, when put into language at the direction of the critic, is able to decipher the meaning that was concealed by ideological distortion. In principle, this would give the critic an *absolute advantage* in any communicative situation, because this knowledge can assure the critic of the validity of his or her critique and indicate the means by which to correct the distortion. Habermas claims that his critical theory provides this meta-hermeneutical knowledge on the basis that the power of reason, though incarnated in language, essentially transcends language.

The dialogical ontology of hermeneutics contests these premises about the nature and limitations of linguistic ability and these perceptions of contextual disadvantage and absolute advantage on the part of the critic. We can point to three claims which Gadamer invokes, explicitly and implicitly, in his response to Habermas which together describe a very different conceptualization of language and the problem of ideology. First, *all* linguistic assertions mean more than they say;[90] second, latent meaning is linguistically accessible and intelligible in *every* instance;[91] and third, our understanding of that latent meaning will *always* be partial, an aspect of it.[92] These claims are central to Gadamer's thesis that language is the horizon of all human understanding. They were first encountered in Chapter One and were invoked in Chapter Three to interpret Gadamer's reply to Strauss; the task here is to show how they address the practical problem of ideology and thereby substantiate Gadamer's defense of hermeneutics and his counter-critique of Habermas.

These claims work together to the following effect. Unspoken and seemingly unintended claims, meanings and knowledge are always present in communication and always exceed our spoken and intended claims, even when we express all that we are aware of. Since we know this about the speech of others and even of ourselves, it is evidence that we, whether ideologue or critic, are not confined (at least not in retrospect) only to what was said, but can know of what was not said but still

---

[90]   "There is no assertion that one can grasp only through the content it presents if one wants to grasp it in its truth. Every assertion is motivated. Every assertion has presuppositions that it does not assert." (Hans-Georg Gadamer, "Was ist Wahrheit?," *Kleine Schriften*, 4 vols. [Tübingen: J. C. B. Mohr, 1967], 1: 54, trans. and quoted by Hoy, *Critical Circle*, 122.)

[91]   "I maintain that the hermeneutical problem is universal and basic for all interhuman experience, both of history and of the present moment, precisely because of the fact that meaning can be experienced even where it is not actually intended." ("On the Scope and Function," 87.)

[92]   "Reflection on a given understanding brings before me something that otherwise happens 'behind my back.' Something—but not everything, for what I have called *wirkungsgeschichtliches Bewußtsein* [the consciousness of historical effects] is inescapably more *being* than *consciousness*, and being is never fully manifest." ("On the Scope and Function," 92. Italics his.)

meant. *Therefore, it is impossible for language to be utterly stymied, much less vanquished by ideological distortion.* This self-reflective character of linguistic being can be suppressed, but there is no ideological distortion or historical contingency which can eliminate it. But *precisely the same rule—* that meaning exceeds what is spoken and intended—applies to all efforts to capture every unsaid thing in spoken form. No theory or contingency will give us a complete knowledge of all our meanings. Our capacity to know about unsaid claims is always limited, even as it is always present.

The hermeneutical viewpoint has premises which are not entirely dissimilar to Habermas', but which already incorporate Habermas' observations in a more profound and comprehensive perspective. Habermas correctly observes that the ideologue means something more than what he or she says. For Gadamer, this points to a universal feature of language and is not due to some defect in language nor to some unassailable power of ideology. To use Habermas' terms, 'ordinary' language is like 'distorted' language in that it too has the character of 'asserting more than it says.' In this sense, 'ordinary' language is already 'distorted' since not all unsaid meaning can be made explicit. Thus 'ordinary' and 'undistorted' language are not fundamentally or essentially different.

This does not mean, however, that all language is as impenetrable as Habermas thinks ideological language is. Hermeneutics takes the opposite position: in principle, ideologically 'distorted' language is as intelligible as 'ordinary' language. Here too, however, the explanation is not completely alien to Habermas' perspective. Habermas posits the transcendental power of reason as the basis by which the critic can be assured of the validity of his or her critique. Gadamer also asserts the possibility of critique but locates it within linguisticality and as something already essential to human being. Gadamer is not averse to calling this possibility a characteristic of reason or reflection; what he objects to is the dichotomy Habermas imposes between language as manipulable and conditional tool, and reason as transcendental and infallible power. Reason and language are rather co-extensive, which does not connote an ultimate powerlessness, as Habermas fears, but a recognition that we are participants in, yet not masters of, the common world which language gives us.[93] Reason does not 'have' a language since language is more than a tool, nor is reason incarnated, as Habermas puts it, in language because language is more than a form. For hermeneutics, reason *is* lin-

---

93 As noted in Chapter Three, Gadamer states that "the speculative character of being that is the ground of hermeneutics extends as universally as does reason and language." (*TM*, 477 [434].)

guistic: the 'language' of reason is language itself.[94] The power of language is never nil, but it is also never infinite in the way Habermas hopes to find in a transcendental reason.

On the basis of these conceptions of language, the hermeneutical perspective opposes Habermas' perceptions of absolute advantage or contextual disadvantage on the part of the critic depending on whether, respectively, the critic has or has not acquired a meta-hermeneutical theory. The difference between the 'said' and 'unsaid' is never so great that it is utterly impossible for ordinary language to bridge it. Consequently the critic is never in a position of complete linguistic disadvantage with regard to an ideological context, and the ideologue is never utterly blinded by ideology as though it were an independent force with total control over human consciousness. If the hermeneutical conception of understanding applies to understanding 'ordinary' language, as Habermas acknowledges, it also applies to understanding 'distorted' language. There is, consequently, no need for an absolute advantage through an extra-linguistic source of knowledge such as Habermas claims to offer in critical theory.

Furthermore, the difference between what is 'said' and 'unsaid,' cannot be absolutely bridged with the systematic and certain assurance Habermas hopes to achieve by recourse to a meta-hermeneutical theory. The 'unsaid' meanings of language are intelligible but not exhaustible, and therefore language not only denies a contextual advantage to the ideologue but also denies an absolute advantage to the critic who wields critical theory (or any theory). And despite the fact that Habermas' emancipatory intent is quite different from the oppressive intent of ideology, it should be noted that Habermas seeks to validate critical theory in the same way as the ideologue seeks to validate his or her ideology: through an absolute advantage over all other points of view. Such an expectation is the source of the danger that critical theory itself will become an ideology.

---

[94] Gadamer's opposition to the idea that knowledge (whether from reason or from "insight") is separable from speaking, is also stated here: "In my work, I have tried to overcome the prejudiced description that all speaking is a secondary experience over against a separate insight and that to speak is a secondary dispensable moment in real communication. Neither of these statements is true. My position is precisely that living communication is always the full phenomenon, the full appearance." ("Discussion on 'Hermeneutics and Social Science,'" *Cultural Hermeneutics* 2 [1975]: 334-335.)

Our account of the hermeneutical view of ideology does not mean, however, that there are no distinguishing features of ideological language or that interpretation of it does not present distinctive problems. Nor does this perspective mean that there is no value in theorizing about the nature of language or the ways to correct linguistic distortions and false agreements. It does mean that theorizing has no extra-linguistic origin or status, and that the aim of such theorizing will be distorted if it fails to recognize its own hermeneutical character. Whatever a theory or a tradition or a philosophy can tell us about good and bad practice, including the practice of communication, that truth is not ahead of language nor located outside our historicity, but in itself demonstrates the linguisticality and historicity of the truth to which it points.

Moreover, this hermeneutical perspective does not imply that every social communication in which one party seeks to penetrate the ideology of another will be successful. Nevertheless, this does not necessarily falsify the hermeneuical claim. Although Gadamer does not pursue this matter, it is obvious that there can be many explanations for why critique may fail to effect change in social thinking or action, not the least of which are the rewards either party may perceive in maintaining a viewpoint that refuses to find a common reality with another. But such problems are often a matter of moral choices. The hermeneutical claim only asserts that the failure to reach true dialogue and communality is not something which can be attributed to some defect, weakness or limitation in our linguisticality. Stated in positive terms, the linguisticality of human being is sufficient for common understanding to occur, and every instance of successful dialogue demonstrates this. This claim also implies that recognition of ideological distortion by the victim or the perpetrator of such a distortion is not something that ideology has the power to fully eliminate. As much as history demonstrates the power of ideology to conceal the true common reality, it also demonstrates that no victim population is ever totally or permanently unaware of this event, showing that the concealment is never completely successful. Theories, then, are not absolute pre-conditions for ideology-critique but aids to it; and every such theory reflects the arrival in self-conscious expression of precisely those things which ideology attempts to conceal: that what the perpetrator of ideology says does not signify a common reality but a desired yet false reality.

In practical life, we may at different times find ourselves in the role of either critic or ideologue. Since the hermeneutical perspective explains how neither the ideologue nor the critic have any final advantage over each other, only dialogue 'from the bottom up' is the appropriate way to engage in practice.[95] And only the awareness that no one is finally advantaged in knowing truth or achieving good can be called the 'advantage' of the one who seeks and practices dialogue. This advantage is not a possession of something, however, but a recognition of the common reality in which both parties participate. This awareness can be nurtured by theories, traditions, or social conditions, but none of these can create this awareness by their own power. It is an awareness which is always available to everyone by virtue of our linguisticality.

Ends and Means in Practice

The preceding interpretation of the hermeneutical assertion—that language is not totally stymied by ideological distortions and therefore does not require us to invoke an extra-linguistic source of knowledge, as Habermas seeks to do—is basically a defensive one. It can be complemented by more affirmative arguments to show how the hermeneutical perspective understands the way by which persons can recover and discover genuine agreement and common reality. To this end, we discuss the significance of Gadamer's concepts of solidarity and rhetoric, and argue that they include and go beyond what is valid in Habermas' concepts of emancipation and critical theory, respectively.

Habermas' objection to a hermeneutical concept of solidarity as a practical principle would be its lack of a critical stance that is committed to emancipation from ideological domination. In fact, solidarity does entail an emancipatory concern and a corresponding critical capacity. Recall that the hermeneutical conceptualization of solidarity is rooted in the intellectual virtues (understanding, judgment or insight, and judging-with) as well as in such moral virtues as friendship and justice: by developing and applying these virtues, we can find solidarity and community with others. As we saw from both Beiner and Gadamer, the three intellectual virtues involve discerning the meaning and reality of persons 'beneath' what appears on the surface and therefore imply a critical capacity. Recall also that Gadamer describes "insight" as the basis for recognizing authentic authority and, by implication, for distinguishing it from false authority. In this context as well, then, the virtue of insight or judgment (*gnome*) involves a critical capacity which necessarily implies

---

[95] "In this game [of language] nobody is above and before all the others; everybody is at the center . . . and thus it is always his turn to be interpreting." ("On the Scope and Function," 88.)

emancipation from false authorities or, in Habermas' terms, from ideology.

However, even if Habermas accepted that such virtues have critical capacities, he would contend that these capacities are contingent on each person and his or her historical context, and are vulnerable to being deceived by ideology; therefore the reliability of this critical capacity of the virtues cannot be guaranteed. It is true that insight and other virtues are aspects of the self and are not infallible instruments, as Habermas regards reason; we cannot be certain, in the way Habermas seeks, that what we think insight tells us is in fact an insight and not a self-deception. From the hermeneutical perspective, however, Habermas' theory is itself located within language and is no less vulnerable to self-deception. What is trusted in Gadamer's concept of "insight" is that the potential for genuine insight is always present and never completely eliminated by ideology. This represents the positive implication of the negative conclusion drawn in the first part of this Constructive Interpretation, that no one has an absolute advantage over another in terms of manipulating language to either prevent insight into ideologically concealed meaning or to step beyond language to transcendental knowledge. This does not mean that linguisticality automatically grants us full and mature insight; it means we always have the possibility of insight within our nature, but it needs to be disciplined and developed and therefore is called a virtue.

The significance of Gadamer's norm goes further. Habermas treats emancipation as the primary or highest goal of practice. Emancipation is indeed an appropriate moral goal in many situations where domination and oppression (and not only of a corporate or social kind) are the basic problem. Beyond the scope of such situations, its value remains, but its importance is not as central because it does not tell us the goal of practice in non-oppressive situations. This limitation appears in Habermas' treatment of social relationships. His theory is satisfied to assert that relationships should be based on free agreement, which primarily means a prior agreement on the conditions under which mutual understanding should be sought so that disagreements are not deceptively resolved to the unfair advantage of one party. If substantive agreement then occurs, the participants may regard this as even better; but at this point Habermas' theory can only remain neutral because it does not include any principle about the substantive nature of good relationships between persons. Moreover, by defining emancipation in terms of free agreement, it tends to become a cause for self-preoccupation on the part of practical agents, and a reason to maintain a perpetual suspicion of any or all claims. The absence of a principle about the substantive nature of good relationships shows emancipation to be inadequate even in regard to Habermas' concern with situations of ideological oppression: emancipation does not tell us toward what new substantive relationship oppressors and oppressed should aim. Thus even in this problem, emancipation cannot be regarded as the primary goal but an aspect of the goal.

In contrast, solidarity by definition is a principle about the substantive nature of good relationships, as is apparent from the kinds of virtues with which Gadamer associates solidarity. Even though these virtues and the acts to which they lead are given minimal definition by Gadamer, it is clear that, taken together, they necessarily link the fulfillment of self to the fulfillment of others, and ultimately of all linguistic beings, in a way emancipation does not. These features of the principle of solidarity also give it a larger scope of applicability to practical human problems that are beyond those which primarily involve oppression or ideological distortions. Problems of dilemma or guilt or grief are examples for which emancipation is not necessarily an adequate, or even a relevant, norm, but for which solidarity is.

This larger scope and more substantive character of solidarity as compared to emancipation is most evidenced by the intellectual virtue of judging-with (*syngnome*). The one who acts in solidarity is not only ready to judge whether the claims of tradition or persons are legitimately authoritative, but is ready to judge-with them and, in effect, to look for *potential* solidarities which the other party does not assert—and perhaps wishes to conceal or fears to acknowledge—but which insight reveals. The notion of potential solidarities is consistent with Gadamer's assertion that communality with others is the "decisive condition and basis" in which we participate even before we make choices. This communality is already apprehended as a latent solidarity when we consciously and socially act to realize new solidarities. Thus, Gadamer's concept of solidarity not only implies the critical capacity that leads to emancipation from ideological oppression, but goes further and points to new possibilities of discovering common reality through the virtues made possible by our linguisticality. Such an emancipation leads *to* genuine communion, and is not merely a departure *from* domination.

From this perspective on solidarity, the compatibility of critique and authority is made more intelligible. To judge-with alleged authorities as well as to judge them takes us beyond asking whether one can know if a claim is a true claim and whether one's acceptance of it is free. This important latter concern is now located within a more comprehensive question put to alleged authorities on which rests the possibilities of both explicit and potential solidarities: 'Do your claims (seek to) judge-with me as I (seek to) judge-with you?' Claims which do not intend to judge-with others but rather essentially judge-for them or judge in their place (and thereby dominate them) are alienating claims and work against solidarity; they do not have any true authority and ought to be critically exposed as such. But claims which do judge-with others (seeking potential solidarity) are simultaneously claims which contain some true understanding and judgment about them; these are claims, then, by which persons gain understanding of themselves, of others, and of their common world. It is on this basis that authority should be granted to the source of those claims. The point of the hermeneutical perspective is that this granting of legitimate authority *can* occur, regardless of how often this granting does or does not occur in history; for Gadamer, it happens more often than we are aware. In any case, the determination of authentic authority is part of the way of life that leads to solidarity because solidarity is built on the shared discernment of these authorities. And since such a determination presupposes a critical capacity, it is necessarily part of the principle of solidarity.

Contrary to Habermas' reading of Gadamer, philosophical hermeneutics in principle imputes to people a greater ability to distinguish insight from illusion, and solidarity from its pretense. Ironically, it is Habermas who views persons as essentially helpless unless they can control and apply the authority of a rationality which is not inherent in human life. But, in fact, this view of reason is an appeal to an ultimate authority which can supersede and judge all other authorities, and it is this ideological reliance on reason which finally proves, in Habermas' own words, to be "blind and slavish."

We turn now to compare Gadamer's concept of rhetoric to Habermas' concept of critical theory. Both Habermas and Gadamer argue that social communication has a central role as a means to the realization of practical good. For Gadamer, such communication is essentially rhetorical and should be guided by the discipline of rhetoric. While he invokes the classical roots of this discipline, his hermeneutical perspective and his concept of solidarity focus on certain features of rhetoric. Rhetoric, as both a discipline and as an activity, does not mean a way of fooling or deceiving someone into accepting something as true, but, as we saw in Gadamer's replies to Habermas, is an activity that participates in knowledge of the good and concretizes it in the act of communication. Rhetoric instructs us through its knowledge and techniques about how to speak, argue and listen so that the possibility of discovering some point of solidarity on which further dialogue can build will be realized.

Rhetorical acts should therefore be guided by the intellectual virtues noted earlier, among which judging-with is especially illuminating when this virtue is taken to mean a readiness to find a common reality with others. From this perspective, rhetorical acts require understanding and insight about the other person so that one's own claim, which is initially foreign to the other, can be presented in a way that will no longer be utterly foreign, but heard and possibly accepted. To engage in rhetoric, then, presupposes that one is continually discerning how one's own claim has some common ground with the other and appropriately reconceiving or modifying the claim so that a common reality and a common claim will appear; this does not mean compromising for the other but judging-with the other. Such a commitment expects the roles of speaker and hearer to be successively reversed, making rhetorical acts part of an ongoing dialogue. Thus rhetoric and solidarity work hand in hand because both depend on the same virtues.

Habermas' objection to this argument for rhetoric, like his objection to solidarity, would be that it takes no account of ideological conditions and linguistic distortions, and even if the speakers do not intend to deceive others, they may already be deceived and so deceive others in their practice. The notion that rhetorical techniques should be guided by virtues would not modify his objection, since he would argue that we cannot count on people to be virtuous if we want to ensure successful communication. Habermas views critical theory, especially its explanation of communicative distortion and communicative competence, as a way of making up for the deficiencies or limitations of rhetoric. This application of critical theory to social communication fundamentally depends on his philosophical claim that its knowledge 'by-passes' language and dialogue, and therefore is presented as a discipline for objectively certifying our understanding of all the meaning carried by speech, but especially for recognizing and penetrating ideology. Consequently, the classical concept of rhetorical discipline, and by implication the virtues on which it is based, are subordinated by Habermas to the prior judgment of critical theory.

The virtues on which successful rhetoric depends are not, of course, always present or sufficiently mature to promise the detection and elimination of ideological distortion. Nor do the skills and techniques of rhetoric absolutely guarantee undistorted agreement. But in order to be consistent with the hermeneutical perspective, this lack of methodologi-cal reliability must be regarded as a 'disadvantage' which also applies to critical theory. Again, the conclusion of the argument offered at the outset of this section applies here: critical theory does not have any final advantage over language such that it can objectively secure the truth or freedom of social communication in a way unavailable to other kinds of linguistic knowledge. Habermas' mistake is to think that the 'limit' of linguistic understanding can be surpassed by his objectivist concept of reason; consequently, critical theory labors under the illusion that it has an advantage over rhetoric. If there are no means by which we can leap ahead of the other person to know what dialogue cannot reveal, or let that knowledge 'do our speaking for us,' as it were, we have no alterna-tive but to put ourselves 'on the line' in our choices about how to engage in social communication. In this respect, critical theory is on an equal footing with rhetoric.

The merits of rhetoric and critical theory as disciplines of social communication can be compared on other grounds by examining the correspondence between means and end in each case, and judging which correspondence best addresses the nature of practice. Since both Habermas and Gadamer conceive of social communication as an impor-tant means to practical good, it can be expected that their conceptions of the manner in which critical theory and rhetoric, respectively, discipline this means will illuminate the way in which they conceive of the end. Habermas' critical theory subordinates the function of classical rhetoric to its own higher authority in order to compensate for the alleged unreli-ability of rhetoric as a discipline of social communication. As such, criti-cal theory is not so much an aid to rhetoric as a 'meta-rhetoric' or a tech-nology for speaking; consequently, social communication is seen primar-ily as a raw material to be technically organized for the purpose of arriv-ing at emancipation and free agreement. The means to practical good, in sum, tend to be technicized by critical theory. This technicizing of means seems to have corresponding appearances in Habermas' conception of the end of practice: true agreement is essentially seen as unanimous agreement of assertions among participants, and emancipation as the ab-sence of restraint or coercion. While these definitions are not irrelevant to practical good, they do not, as noted earlier, identify very much of the substantive character or content of the practical good. As a result, practi-cal good tends to take on the character of a technical product, and this, it seems, is related to his technical approach to social communication.

The hermeneutical use of rhetoric, however, is conceived as a practical discipline and not an essentially technical one: its technical knowledge is subordinated to and guided by the virtues which characterize the goal of solidarity so that these virtues also guide the way social communication is engaged in as a means to that goal. In fact, as noted earlier, Gadamer implicitly asserts in "What is Practice?" that solidarity must characterize the means (it is the "decisive condition and basis" of "social reason") as well as the end of practice. (And since other implied objections to critical theory appear in the context of this particular assertion, we may suppose that it is also designed in part to oppose Habermas' position.) Unlike critical theory, then, rhetoric presupposes that to reach practical goals we must use genuinely practical means.

While this axiom frustrates a desire to arrive at practical goals in a more efficient way than rhetoric can provide, it points to a further, common-sense implication of the correspondence between means and ends: there is no part of the means to practical good which is more reliable than (nor essentially different from) the end of good practice. Habermas implicitly contradicts this principle when he avoids revealing the substantive content of the end of practice and settles for a truncated, more technical version of it, the only version which is as reliable as his chosen means—that is, social communication which has been technically disciplined by critical theory. In other words, by trying to arrive at practical good with more certainty and more speed than rhetoric allows, the application of critical theory tends to technicize the means of practice, and therefore implicitly technicize the ends of practice as well; in doing so it fails to grapple with the full practical character of either the means or ends.[96]

---

[96] The incompatibility here between the uncertainty of achieving practical ends and the certainty of technical means is evident in the way Habermas asserts, on the one hand, that "the conditions of empirical conversation are obviously not, or at least often not identical with those of the ideal speaking-situation," but on the other hand, goes on to say that "the formal anticipation of idealized discussion (as a form of life to be realized in the future?) [*sic*] guarantees the final supporting, contra-factual consensus which must previously connect the potential speakers/hearers." To assert the actual elusiveness of the ideal conversation, and therefore of free agreement, is inconsistent with asserting that we have objectively certain grounds for anticipating the realization of this ideal. Of course there is no inconsistency if practical good is conceived as a matter of technical causation; however, this is not only false, it also misperceives "conditions" as causes which effect the good. Nevertheless, Habermas maintains that "linguistic agreement is . . . both: it is anticipated, but as an anticipated basis it is also real." (Jürgen Habermas, "Summation and Response," *Continuum* 8 [1970]: 132.) This appears to be similar to Gadamer's position, although for Gadamer

The value of critical theory, in terms of its knowledge about the conditions of distorted communication and about rules and techniques for undistorted communication, is real but more modest than Habermas supposes. Its value can be properly estimated only when its knowledge is subordinated to a discipline which presupposes the correspondence of means and end in practice, as does rhetoric. The apparent weakness of rhetoric is its 'advantage': it knows that its form of disciplining social communication does not have any special capacity to achieve practical good that is not already part of that goal, and it recognizes that alternatives which claim such a capacity are illusory.

Incompatibility of Philosophical Hermeneutics and Critical Theory

It might be asked if these arguments against Habermas reflect a misunderstanding of his theoretical position. Critical theory may not be so different from philosophical hermeneutics as has been implied. In fact, perhaps critical theory is compatible with it, or even supersedes it, by incorporating the primary merits of the hermeneutical perspective while excluding its problematic aims or implications. Whether or not this is Habermas' own view, it is certainly one which is encountered, in various versions, among other writers who seem to think critical theory is a better resource for social ethical inquiry than philosophical hermeneutics.

For his part, Habermas' objection to philosophical hermeneutics is by no means total, and there are a number of areas of apparent agreement with Gadamer. In one place, for example, Habermas catalogs how Gadamer's philosophy corrects scientific prejudices, reminds science of its limitations, and mediates its knowledge to social life.[97] And there are a number of other instances, some of which were quoted earlier in this section, where he qualifies his critique of Gadamer in order to make clear that he does not presume it possible to stand utterly outside of language, nor to acquire absolute knowledge. For example, after defending his

---

agreement is already real in the act of mutual understanding; its possibility is not due to our power of anticipation but to the nature of the world itself, to which our anticipations are responses.

[97] As Habermas enumerates them, the merits of philosophical hermeneutics are: it "destroys the objectivist self-understanding" of the human sciences with its argument for the effective historical connection between subjects and objects; it "reminds the social sciences" of the presence of prejudgments in every inquiry; it reminds the natural sciences that "natural language represents the 'last' metalanguage for all theories expressed in formal language"; and finally, it guides us in "the translation of important scientific information into the language of the social life world." (Habermas, "Hermeneutic Claim," 186-187.)

concept of critical reflection, he adds this toward the end of his second essay dealing with Gadamer:

> It is, of course, true that criticism is always tied to the context of tradition which it reflects. Gadamer's hermeneutic reservations are justified against monological self-certainty which merely arrogates to itself the title of critique. There is no validation of depth-hermeneutical interpretation outside of the self-reflection of all participants that is successfully achieved in a dialogue.[98]

Similarly, Habermas emphasizes that while he finds Gadamer's viewpoint to be inadequate, he is not seeking an Hegelian alternative:

> Reflection can, to be sure, no longer reach beyond itself to an absolute consciousness, which it then pretends to be. The way to absolute idealism is barred to a transcendental consciousness that is hermeneutically broken and plunged back into the contingent complex of traditions. But must it for that reason remain struck on the path of a relative idealism?[99]

These and other passages suggest that Habermas intends his critical theory to be a theory that acknowledges the hermeneutical insight about the effective history of knowledge and the impossibility of absolute knowledge. Such qualifications are probably appealing to those who would like to find some middle position between Habermas and Gadamer, or prefer to soften Habermas' claims with Gadamer's.

But if Habermas' philosophical critique does not represent a position so different from hermeneutics, what is the aim of his critique? Habermas' statement at the close of his second and last major essay on Gadamer sheds some light on this:

> In the present conditions it may be more urgent to indicate the limits of the false claim to universality made by criticism rather than that of the hermeneutic claim to universality. Where the dispute about the grounds of justification is concerned, however, it is necessary to critically examine the latter claim, too.[100]

Considering that Habermas' attention to Gadamer is primarily critical, this is a surprising concession to the contemporary relevance and value of Gadamer's claim. He does not reject but shares Gadamer's critique of "false" criticism, that is, of criticism which arrogates a false universality

---

[98] Habermas, "Hermeneutic Claim," 209.

[99] Habermas, "Review" 359-360.

[100] Habermas, "Hermeneutic Claim," 209.

to itself. The point of the second sentence appears to be that he also opposes Gadamer's claim to the universality of hermeneutics since, according to Habermas, Gadamer does not take adequate consideration of how language itself can be used to distort the world for us, and thus cannot be given the universality Gadamer ascribes to it. No doubt Habermas believes that his own constructive project is adequately qualified by statements such as those noted above so that his own theory is not one of those "false claims" for the universality of critique to which he refers.

If Habermas actually means to argue that we can engage in a more intensive or more thorough reflection—though still finite and linguistic—on all forms of human understanding than Gadamer has yet explored, then perhaps he is still within the scope of an intrinsic critique of philosophical hermeneutics. In this case, he is engaged in a dispute with Gadamer over what finite reflection actually tells us about the social world. This would not be a refutation of Gadamer's philosophy, but might be a demand for a more adequate demonstration of it or an invitation to pursue its implications. But, in fact, we are dealing with a claim of a different order: as shown at the beginning of this chapter, his critique is predicated on the notion that hermeneutics is inadequate to deal with distorted language. He goes further and claims that we have access to non-linguistic or non-hermeneutical knowledge. This contradicts a basic thesis of philosophical hermeneutics and thus demonstrates that he is not engaged in an intrinsic critique.

It is not possible to reconcile Habermas' selective appreciation of Gadamer's hermeneutical perspective with the over-all intent of critical theory, which is to go beyond hermeneutics, not merely in its investigations into social life, but in terms of acquiring some kind of transcendental foundation which philosophically can assure us of the validity of critique. Even a charitable interpretation must point in this direction. Habermas asserts that criticism is "tied" to its traditional context, and at the same time also asserts that critique "must . . . not be tied to the radius of convictions existing within a tradition."[101] As we saw Gadamer assert earlier in this chapter, it is quite appropriate to view our relationship to tradition as one which is critical and creative as well as conservative; this could be taken to mean that in some sense we are indeed both tied to, and not tied to, tradition. But what Habermas intends goes beyond this and is self-contradictory: he wants both to retain his connection to tradition, lest his objective become an "absolute consciousness," and to de-

---

101 Habermas, "Hermeneutic Claim," 208.

clare and verify his absolute freedom from tradition and from any of its claims. Habermas' statement, quoted above, that dialogue alone can validate an interpretation derived from critical theory, is close to Gadamer's position; but it is incompatible with his own claim that critical theory provides knowledge which "transcend[s] the limits of the hermeneutical understanding of meaning."[102] Thus, if there are indeed elements of philosophical hermeneutics being affirmed by critical theory, they are incompatible with the basic purpose of that theory.

The difference in the aims of these writers is exemplified in Habermas' response to Gadamer's concept of application. He appears to understand and appreciate Gadamer's insight that (in Habermas' words) "interpretation is realized in the application itself." But he goes on to say this: "Application is problematic and inseparable from interpretation wherever a transcendental framework is not yet established once and for all but is undergoing transformation and must be decided ad hoc."[103] This implies that once a "transcendental framework" is thoroughly worked out, application will no longer need to be troubled by its present dependence on interpretation and its uncertainties but will have clear, unambiguous rules to follow. If Habermas thinks application can overcome its interpretive (hermeneutical) character, the implication, contrary to his qualifications, is that he thinks it possible to be 'ahead' of our historicity. This direction of thought is fundamentally different from the hermeneutical claim that application and interpretation are inseparable, not just occasionally but essentially, due to the historicity of all human knowing and doing.

A theory of truth that claims to 'by-pass' natural language cannot also claim to be a theory truly subject to immanent dialogue. It cannot claim to provide methodological certainty and yet also claim to be appropriate to ordinary language which unfolds without any such certainty. It cannot remain linguistically 'sane' and at the same time perpetually suspect that every claim is distorted. Habermas may wish to have it both ways but he cannot, and it is clear on which side of these alternatives he puts greater weight: on the theoretical products of critical reason as a means to guarantee the 'freedom' of truth from the historicity of truth.

There may be some intention on Habermas' part to fulfill the promise of Gadamer's hermeneutics for practical philosophy in a way which

---

[102] Habermas, "Hermeneutic Claim," 180.
[103] Habermas, "Review," 355.

Gadamer has not done. But if so, his manner of doing so is neither consistent with philosophical hermeneutics nor does it require us to modify our criticism of critical theory. Critical theory has significant value, which lies most of all in its reminder of the power of ideology and its analyses of conditions which abet or dissolve that power. However, while these merits draw on societal inquiries not made by Gadamer, they do not reveal non-hermeneutical knowledge and they inadequately deal with the non-objectifiable and essentially hermeneutical conditions of human practice.

In the final analysis, both Habermas and Gadamer view their respective projects as incompatible because of Gadamer's ontological claim. Habermas wishes to retain hermeneutics, but without this ontology since it contradicts his own meta-hermeneutical claim for reason. He wishes instead to use what would be left of hermeneutics as an essentially historicist tool in critical theory. For his part, Gadamer would doubtless consider a critical theory which discards his ontology to be precisely what Habermas calls a "false claim to the universality made by criticism." The dialogical ontology is not an appendage to philosophical hermeneutics but its basis, and therefore, for Gadamer, no theoretical project which disposes of this ontology can be considered compatible with his philosophy.

On the theoretical level, the source of Habermas' latent ambivalence is a desire to affirm two incompatible perspectives: a hermeneutically-oriented historicism and a scientifically-oriented objectivism. The position proposed by Gadamer is more parsimonious: we do not need to choose between historicism and objectivism since, in the dialogical ontology, the truth in each requires the truth in the other. Nor do we need to let the possibilities of effective critique or of genuine agreement be held hostage to a false antithesis between transcendental reason and contingent language. In view of the alternative offered by Gadamer, it appears that Habermas opposes tradition to reason not so much because of his correct concern to penetrate ideology, but because he cannot think past the Enlightenment's dichotomy between reason and tradition.

There is also a more practical and existential level to Habermas' ambivalence: he seems to be unsure of the safety of acknowledging the finitude of human thought and practice. He only conceives of this finitude as an imprisonment, which is reflected in his concepts of language and tradition as closed circles of ultimately random—or at least untrustworthy—activity. The role and status given to reason and emancipation indicate that he places on them the hope of escaping this imprisonment.

By seeking something with absolute power by which to levitate human knowledge above human finitude, Habermas fails to grasp the full dimensions of practice. By contrast, the depiction of human historicity presented by Gadamer's dialogical ontology suggests that confidence in the possibility of good practice is not misplaced or illusory. Human finitude is not an imprisonment but a concomitant of our participation in the whole of being, and it is in the hermeneutical concept of historicity that this union of finitude and participation is reflected. Consequently, as we approach the tasks of human practice, we are neither so helpless in actuality, nor so transcendentally powerful in potentiality, as Habermas implies.

This conception of human life does not mean that hermeneutics necessarily minimizes the risks, temptations and dangers which accompany the practical exercise of language, nor that it supposes good practice to be inevitable because our linguisticality is inescapable. It does mean that no theory or technique can filter out all the dangers or guarantee the success of practice, and it is in this sense that the possibilities of good practice open to humans through language and tradition are indeed contextually relative to one's particular language and tradition. But such particular languages and traditions constitute the scenes in which our participation in the whole of reality become evident to us and by which we recognize our capacity for genuinely practical and moral choices. Thus, the very historicity of life, which falsifies our efforts to guarantee good practice, is the same historicity which makes it possible to bridge alienations and penetrate ideological domination, and to discover common reality with others.

CONCLUSIONS FROM GADAMER'S DEBATES WITH STRAUSS AND HABERMAS

We are now in a position to survey the results of this and the previous chapter as they bear on our inquiry into the possibility of a hermeneutical philosophy of practice that has theoretical viability and practical relevance. The general objection to this possibility, as observed in the beginning of Chapter Three, is that philosophical hermeneutics is a relativist philosophy and therefore irrelevant to a philosophy about and for moral practice. This general charge, conceived in two forms—inherent relativism, represented by Strauss' critique, and contextual relativism, represented by Habermas' critique—implied the practical dangers of nihilism and ideologism, respectively. We hypothesized that if the integrity of this philosophy could be defended against their critiques, we would have confirmed in principle that philosophical hermeneutics has relevance for our understanding of practice and supports confidence in the possibility of good practice. The results of our analyses of these debates can be recapitulated with reference to this objective.

The Theoretical Viability of a Hermeneutical Philosophy of Practice

Chapter Three considered whether philosophical hermeneutics is an *inherently* relativist philosophy by examining its response to Strauss' critique that its historicist character inevitably dissolves all truth and contributes to nihilism in practical life. Strauss thinks Gadamer's philosophy can be salvaged if its relativism is renounced and the remainder is made an instrument of political philosophy and his own natural law absolutism. When Gadamer's responses to Strauss are interpreted in light of his ontology, as was done in our concluding section of Chapter Three, philosophical hermeneutics shows that it does not have an objectivist concept of truth, as does radical historicism. On the basis of that ontology, hermeneutics finds truth of a different kind in history. Therefore, hermeneutics is not inherently relativist in the sense that it does not dissolve all truth into arbitrary claims and does not contribute to a nihilistic outlook.

However, philosophical hermeneutics is relativist in another sense. Humans can have true knowledge and make universal moral claims, but these reflect the participation of finite, historical beings in the very realities to which such knowledge and claims point. Truth, then, is relative in the sense that it is relational, and therefore exists in a dynamic historical process as persons relate to each other, the world, the past, and to all of reality which humans can understand. Thus, hermeneutical relativism affirms the genuineness of truth in a manner consistent with human historicity. *Indeed, true understanding is possible, not in spite of our historicity but with it and through it.*

The main body of the present chapter considered whether philosophical hermeneutics is a *contextually* relativist philosophy by examining its response to Habermas' charge that it has no critical capacity toward ideological distortion of social communication and therefore inevitably abets the power of the prevailing ideology. Habermas aims to rectify hermeneutics by discarding its ontology and making it the instrument of a theory of transcendental reason which has the universality he thinks Gadamer's theory lacks, and which can be relied on to expose ideological distortion and secure free agreement. Gadamer's responses to this critique show that the hermeneutical perspective already presupposes that human linguisticality includes a critical capacity. Whatever emancipatory capacity critical theory or any theory has is itself a demonstration of the hermeneutical claim for the universality of language. Therefore, philosophical hermeneutics does not take a contextually relativist position on the possibilities of good practice in that it is neither an apologist for traditionalism nor does it contribute to ideological domination.

But hermeneutics does affirm the contextual relativity of practice in a different sense. Since knowledge is always linguistic, there is no transcendental reason which gives the critic a final advantage in objectively securing critique or emancipation. And because language already has critical and emancipatory potentiality, the ideologue too has no final advantage and there is no need to invoke a transcendental reason in order to effect critique and emancipation. This means that the possibilities of good speaking, and of good practice in general, are indeed relative to—in relation to—our social and traditional context. But this relativity indicates the true nature of practice to be not a technique for reaching an unrelated end, but an act in which the means must express the end itself. From the hermeneutical perspective, this is possible because language is not just a tool but the ontological basis of the participation of finite beings in the whole of reality. As a result, critique and emancipation find their proper meaning within the discipline of rhetoric and the general norm of solidarity, respectively. Thus, philosophical hermeneutics affirms the human capacity for practice in a way that is consistent with human linguisticality. *In fact, good practice is always possible, not in spite of our linguistic nature, but with it and through it.*

Philosophical hermeneutics, then, is not relativist in either of the original forms of this criticism. Yet the ways in which it is relativist are part of an account of the nature of human understanding and practice which is superior to the perspectives presented by Strauss and Habermas. The consequence of this account is that true understanding of the world and of good practice in it are indeed possible. Thus the view that philosophical hermeneutics, due to its relativist character, is irrelevant to a philosophy about and for practice, is erroneous. It can, in fact, provide the theoretical framework of a practical philosophy. But the nature of philosophical hermeneutics, as it has been explored in the course of these two debates, does more than this. *Its response in each debate, and the synoptic result of these two debates, have the effect of making an implicit argument for the compatibility of historicity and good.* The affirmation of this compatibility rests on the way both terms are redefined in the course of these debates by the ontology of philosophical hermeneutics. The historicity of human knowledge is not seen as a barrier to truth and moral understanding but as their scene and even as part of them. Good is not viewed as inaccessible to humans, but as something which is always possible for finite beings to realize and receive within their historical scene. Thus, goodness does not necessarily elude historical beings, and their historicity does not inevitably preclude its occurrence. Put positively, this means that to be full and effective practical agents, our nature is not lacking anything.

Relevance to the Dilemma of Modern Practice

The compatibility of historicity and good was postulated at the outset of this work as a necessary condition for the restoration of confidence in the possibility of effecting good practice in our time. The argument for this compatibility, which we have reconstructed from Gadamer's work and have interpretively amended, bears directly on the way many in our age view moral judgment as being caught in an unresolvable dilemma. Modern consciousness presupposes the importance of a rigorous distinction between fact and value, which is reproduced in the alleged incompatibility of judgments about historical truth and judgments about moral good. This consciousness tells us that we can claim either to know true facts (which are intersubjectively demonstrable) or to affirm good values (in private subjective claims which no one can either confirm or refute), or that we can do both but only in separate 'compartments' of consciousness. In effect, according to the perspective of this 'incompatibility thesis,' we do not live in a single reality, but in two mutually incomprehensible realities.

From this perspective, we have no way of confirming the reliability and validity of any practical ends or means, and as a result modern confidence in the possibility of good practice has faltered and atrophied. The question—which our age hopes to answer affirmatively, yet fears is impossible—can be put in two ways: Can we truly understand history without our efforts resulting, as it seems they must, in value neutrality or ethical relativism? Can we affirm and act upon some idea of good, some tradition of morality, without our affirmations becoming, as seems so inevitable, not only dogmatically narrow but ideologically manipulated, and ignorant of the historical contingency of all affirmations of what is good?

These are questions which in our age and place seem doomed to negative conclusions. From the perspective of the incompatibility thesis, every affirmation of a good is perceived to be either arrogant or cowardly or both, and always suspect of intellectual dishonesty; every acknowledgement of our historical contingency seems unavoidably true yet also lifeless and empty. We are pulled back and forth between the perceived security of submerging ourselves in a moral authority, and the emancipating yet fruitless realization of the apparent arrogance of every such authority. Finding these options unacceptable, we gradually despair of finding true values. Thus the dilemma which characterizes the modern attitude toward practice must either be solved or transformed, but these are precisely the options which appear impossible. Our inner yearning for the coherence of truth and good is precisely what the modern consciousness presumes is most illusory and dangerous, and so it is no wonder that on this basis there is no possible way to conceive reality as housing both true knowledge and real good.

With its argument for the compatibility of historicity and good, philosophical hermeneutics provides a perspective which in principle resolves the philosophical dilemma of modern practice. It does so by offering concepts of historical understanding and practical good which do not signify incompatible parts of reality. And from that perspective, the stark alternatives with which we are presented in our age—fact or value, objectivism or relativism about value, and ideologism or nihilism about practice—are false choices. By implication, the terms of these choices are not really open to the world they seek to explain. To use Gadamer's notion, they are the results not of open questions but of somehow slanted ones. Yet the new concepts offered by hermeneutics are still connected to the language of our age of historical consciousness.

If a practical philosophy is to be intellectually persuasive today, it must be able to speak in this language. But if it is to constructively address the modern dilemma of practice and be relevant to its existential dimension, it must also be able to take us beyond the scientific and historicist prejudices which create that dilemma. Philosophical hermeneutics achieves both of these aims, and thus makes the idea of practical philosophy viable and supports confidence in the possibility of good practice. The case for the relevance of philosophical hermeneutics to practice is made, not by demonstrating it through the presuppositions of the modern dilemma, but in terms of an old yet new conception of practice: the 'fusion of horizons' between Aristotle's practical philosophy and the contemporary experience of historical contingency.[104]

Comparisons: Strauss, Habermas, Gadamer

These conclusions can be meaningfully restated by briefly noting some significant comparisons among the three authors on which these last two chapters have dwelled. On one level of comparison, Gadamer appears to take positions that lie between Strauss and Habermas. Ironically, against each of these interlocutors Gadamer is defending something associated with the other critic; however, he defends it in his own way, on a different basis. Against Strauss' natural law absolutism, Gadamer defends the freedom of practical reason, as Habermas wishes to do; but Gadamer defends the freedom of reason *in* historicity, not *from*

---

[104] Observations by Antonio Da Re about the contemporary significance of Gadamer's efforts to effect this fusion elaborate upon the conclusions we have drawn:

> [Gadamer's] teaching about *phronesis* is a call to responsibility, to dedication, which is quite urgent in our time because we are deeply affected by the destruction of so many idols and myths, by the crisis of the established order. Among such instabilities, the temptation to be rid of such responsibilities and close off ourselves, perhaps in the name of the "social context," is very strong. Instead, the personal choice, the exercising of concrete rationality according to the way of *phronesis*, becomes an invitation not to hide ourselves in anonymity, an invitation made even more urgent by the uncertainty of life itself. . . .
>
> Gadamer has gone beyond Nietzsche's relativistic negativism, beyond the disheartening self-limitation of Wittgenstein's silence, and beyond a resigned waiting because, according to Heidegger's pronouncement, "only a god can save us": positions in contemporary thought which all have discredited ethical reflection. Gadamer has courageously rehabilitated the ancient knowledge of mankind, who, with humility and trust, tried to clarify for themselves the meaning of existence and reality, asking support from the fragile but irreplaceable light of reason. (Antonio Da Re, *L'ermeneutica di Gadamer e la Filosofia Pratica* [Rimini, Italy: Maggioli, 1982] 137,138; passage translated by Cristina Puglisi Kamitsuka.)

Da Re's short book is one of only a few sources known to this author to take up several of the themes on which the present work is also focussed. He is both drawn by the contemporary relevance of Gadamer's viewpoint and sensitive to its limitations (for the latter, see, eg., 133 in his book, and here see Chapter Five note 14, below). A better understanding and more critical study by English speakers of Da Re's contribution must, however, await its complete translation.

historicity, and he defends its capacity for infinite dialogue, not its monological authority. And against Habermas' critical theory, Gadamer defends our connection with tradition, as Strauss wishes to do; yet
Gadamer defends not the static authority of a completed or long-past
tradition but defends tradition as a moving horizon that is shaped
through our own creativity.

Another contrast can also be perceived. Strauss espouses an unhistor-
ical belief in true values, a belief which is not present in Habermas,
whose values, for all their transcendental character, are viewed as functions of the needs of evolving organisms. However, on the one hand,
Strauss' avoidance of this kind of historicism does not bring him any
closer to substantiating certain knowledge of the content of the absolute
values he affirms. And on the other hand, Habermas takes for granted on
historicist grounds the failure of Strauss to absolutely know that content.
But despite this supposed historical honesty, it does not prevent
Habermas from trying to resurrect a kind of absolute truth in the shape
of formal objective criteria which implicitly must transcend and finally
contradict those historicist conclusions. In this respect Habermas is one
of those radical historicists, characterized in Chapter Three, who attempt
to formulate a theory of truth consistent with historicity, yet nevertheless
end up trying to by-pass that very historicity. It seems, then, that Strauss'
allegation of inconsistency, though directed at Gadamer, finds a better
target in Habermas. By contrast, philosophical hermeneutics moves in a
different direction from both the subject-object dichotomy which plagues
Strauss' classicism in an historicist age, and from the form-content dichotomy that hobbles Habermas' understanding of historicity in his pursuit of a different basis for true values. Gadamer's ontology provides an
account of historicity which is not hostage to historicism, and an account
of practical good which is not distorted by a rationalist objectivism.

Finally, the differences between Strauss and Habermas are overshad-
owed by a revealing commonality which contrasts with Gadamer's position. When viewed from Gadamer's discussion of language, both Strauss
and Habermas erroneously tend to see and use language only instrumentally, as a means to acquiring knowledge which objectively corresponds
to reality 'out there'. Consequently, both find Gadamer's ontology to be
the basic source of the limitation they see in philosophical hermeneutics,
although neither appears to have struggled to understand the vision of
language which Gadamer presents as the pinnacle of *Truth and Method*.
One basic consequence of this misconception of Gadamer's practical
philosophy is that they seem unable to fully comprehend what Gadamer

means by application, for in this the sum of Gadamer's ontology is visible.

Because this ontology is itself at the edge of propositional comprehensibility, we should read it as Gadamer intends it: as a speculative description of reality. If this or any such language is instead viewed as an instrumental object apart from the world it embodies, then it becomes a barrier, interminably frustrating efforts to get 'past' it to the real world. For each effort to get beyond language is itself a linguistic effort, making hopeless a venture one nevertheless feels compelled to continue; the language 'game' remains a vicious circle. Of course, moments occur and scenes appear in which the circle seems to be no longer vicious. Such successes in language occur when we arrive at a speculative instead of a propositional mode, when the circle of understanding has not been broken but allowed to fulfill its nature. We then find ourselves 'at rest,' not because our words correctly correspond to the reality 'out there,' but because our finite words have conveyed us to an encounter with something which offers infinite presentations of itself. In such moments we have not required of a word that it give us that infinity, nor have we demanded control over that world; rather, we have found what Gadamer calls concretization, in which the word and the reality are one and yet not one. World becomes word, and simultaneously word disappears into world.

FIVE

Conception of
a Hermeneutical Philosophy of Practice

The fundamental purpose of practical philosophy, as Gadamer describes it, is to help our practical judgments and actions by providing a 'target' of what is good. "As the science of the good in human life, it [practical philosophy] promotes that good itself."[1] The preceding two chapters had a primarily defensive purpose: by addressing the criticisms of Strauss and Habermas, they substantiated the idea that philosophical hermeneutics has relevance to practice and to practical judgment. In the course of this demonstration, these chapters also had a constructive effect: at different points and on different levels of reflection, the relevance of hermeneutics for practice had substantive implications for targeting what is good, as in Gadamer's references to solidarity, dialogue and rhetoric.

Of course, these implications do not go very far to identify what the target of practice should be. There is, in general, a sense that something is missing, a sense that we have not yet heard from Gadamer nor been able to infer from his work the kind of statements that we might expect to find in a practical philosophy. These apparent lacunae involve a level of concreteness—of further targeting what is good—to which Strauss and Habermas give greater attention than Gadamer in their respective projects. Their critiques of Gadamer are, in part, challenges to Gadamer to give these levels of reflection a similar degree of attention. The appar-

---

[1] RAS, 118.

ent need to address this level of normative specificity suggests that we lack clarity about the functions of practical philosophy from the hermeneutical perspective. However, contrary to the supposition of many supporters as well as critics of philosophical hermeneutics, it is not necessary for Gadamer to jettison any basic principles of his philosophy in order to adequately identify these functions, and, together with other resources, to adequately fulfill them. In fact, the principles of philosophical hermeneutics itself point to—and even require—the formation of a more complete practical philosophy, of a more visible and attracting target.

In this chapter, we use an fundamental principle of philosophical hermeneutics—the reciprocity of theory and practice—as the primary basis for an original formulation of the functions of such a philosophy. From that perspective, we then proceed to critically examine more of Gadamer's practical philosophy, including texts which have received little attention until now in the English language. Our aim is to see whether or to what extent Gadamer's elaboration of practical philosophy fulfills the functions that such a philosophy ought to address. In this way, the critical and constructive arguments in this and the final chapter will be intrinsic to philosophical hermeneutics, and unlike the basically extrinsic critiques presented by Strauss and Habermas. We will see that Gadamer's nascent practical philosophy falls short of adequately identifying, much less fulfilling, the functions of practical philosophy. This will not be a great surprise to those familiar with his work. But it is important to make this assessment for three reasons: first, no general survey of this kind has yet been done; second, the criteria of assessment to be applied are, for the first time, derived from Gadamer's own words; and third, what he does say contains some provocative surprises after all, and deserves attention here and suggests avenues for future work.

THE RECIPROCITY OF THEORY AND PRACTICE AND THE FUNCTIONS OF PRACTICAL PHILOSOPHY

Gadamer asserts that the authentic meanings of theory and practice are rooted in their classical definitions and implied reciprocity. In their original Greek meanings, theory is an account of what is unchanging and unconditional, while practice consists of choices in concrete life, which is constantly changing and always conditional. To speak of a reciprocity between these human activities means many things, but foremost and most simply, it means that theory and practice shape, and so are also shaped by, each other.

Gadamer on the Reciprocity of Theory and Practice

In several places, however, Gadamer does refer explicitly to the reciprocity of theory and practice, clearly indicating that it is a central principle of philosophical hermeneutics. It is very significant that, as the following passage demonstrates, he identifies recognition of this reciprocity as the primary affinity between philosophical hermeneutics and Aristotle's practical philosophy:

> The great tradition of practical philosophy lives on in a hermeneutics that becomes aware of its philosophic implications, so we have recourse to this tradition about which we have spoken. In both cases, we have the same mutual implication between theoretical interest and practical action. Aristotle thought this issue through with complete lucidity in his ethics. For one to dedicate one's life to theoretic interests presupposes the virtue of *phronesis*. This in no way restricts the primacy of theory or of an interest in the pure desire to know. The idea of theory is and remains the exclusion of every interest in mere utility, whether on the part of the individual, the group, or the society as a whole. On the other hand, the primacy of "practice" is undeniable. Aristotle was insightful enough to acknowledge the reciprocity between theory and practice.[2]

Gadamer speaks here of both the primacy of theory and the primacy of practice. This apparent contradiction is better explained by describing the relationship between theory and practice as the "mutual implication between theoretical interest and practical action." The meanings implied by this reference to "mutual implication" are indicated by two phrases in other essays: in one, he says that *"theoria* is itself a practice,"[3] and in the other, he reminds us that genuine practice is "characterized by that very possibility of human behavior which we call 'theoretical.'"[4]

The reciprocity of theory and practice is also indicated in Gadamer's interpretation of the equality of the two classical ideals of practical life and theoretical life:

> One may not absolutize the priority given to the ideal of theoretical life over the ideal of practical-political life; Aristotle knows just as well as Plato that for human beings precisely this possibility of the theoretical life is limited and conditional. Human beings cannot devote themselves persistently and

---

2 *RAS*, 111.
3 *RAS*, 90.

uninterruptedly to thought's pure seeing for precisely the reason that their nature is composite. Hence, viewed from the perspective of practical philosophy, the relationship of the two ideals of life is not such that the complete happiness of practical life would not be something supreme too. To be sure, Aristotle calls this happiness a *deuteros*, that is, a second best. But this too is something best, that is, a fulfillment of eudaimonia (happiness). The fulfillment in purely theoretical existence is, after all, not the full bliss of the gods, since it is a limited fulfillment for human beings. The happiness of nous is in a certain sense separate (*kechorismene*)—beyond all comparison. And precisely for this reason the practical happiness of human beings is not second rank, rather precisely what has been apportioned to them.[5]

Gadamer makes this point with greater existential directness in his interview with Ernest Fortin:

> We are mortals and not gods. If we were gods, the question could be posed as an alternative. Unfortunately, we do not have that choice . . . We must take both lives into account. The characterization of the practical life as the second best life in the Aristotelian scheme means only that the theoretical life would be fine if we were gods; but we are not . . . Ours is a fundamentally and inescapably hermeneutical situation with which we have to come to terms via a mediation of the practical problems of politics and society with the theoretical life.[6]

The principle of reciprocity between theory and practice which is either described or implied by each of these quotations is new to our investigation in this explicit form, but the spirit of this principle has been evident in all of our chapters. Gadamer's confidence—that the dialectic of word and world cannot be terminated, that the dialogue between oppressor and victim will not spin into chaos—rests in turn on a trust that our cosmos, in which we ourselves oscillate between theory and practice, is still a cosmos with reason. It could be said that hermeneutical consciousness is an awareness of this reciprocity—and equally an awareness of the finitude of our awareness of that reciprocity, by virtue of our location in it. Furthermore, the same awareness capitalizes on the possibilities of each, and avoids the dangers of absolutizing one at the expense of the other. Theory which is not aware of its origins in practice inevitably dictates our efforts to do good and disregards both our freedom and our responsibility to make choices; practice which forgets its

---

4 "Theory, Technology, Practice," 544.
5 *Idea of the Good*, 176-177.
6 "Gadamer on Strauss," 12-13.

bond with theoretical reasoning devolves into unguided choices and arbitrary whims.[7]

The Reflective and Participatory Functions

An original conception of how practical philosophy fulfills the purpose of targeting what is good can now be unfolded through reference to the reciprocity between theory and practice. Gadamer makes a statement about the reciprocity of theory and practice which is particularly germane to this objective:

> In practical matters the general hermeneutical task which figures in all instances, i.e., of concretizing general knowledge, always implies the opposite task of generalizing something concrete.[8]

We have seen him frequently emphasize the process of judgment which moves from the universal to the particular, a movement he calls *application* or concretization. But here is one of the comparatively few locations where he explicitly draws our attention to the "opposite task," the task of "generalizing something concrete," of moving from particular to universal. Gadamer makes a similar point in another text: "As far as hermeneutics is concerned it is quite to the point to confront the separation of theory from practice entailed in the modern notion[s] of theoretical science and practical-technical application with an idea of knowledge that has taken the opposite path leading from practice toward making it aware of itself."[9] We will call this activity *articulation*.

In his "Afterword" to *Truth and Method*, the significance and dignity of this opposite task is further reinforced: "What emerges from the background of the great tradition of practical (and political) philosophy reaching from Aristotle to the turn of the nineteenth century is that practice represents an independent contribution to knowledge. Here the concrete particular proves to be not only the starting point but also a continuing determination of the content of the universal." On the same page

---

7 Of course, we cannot pretend to have captured the real nature of theory and practice in this principle of their reciprocity. There is a sense in which this talk of 'theory-practice reciprocity' is simply intended to remind ourselves of our tendency to exaggerate one at the expense of the other in our quest for certainty and control. We should not, in turn, make talk of this reciprocity one more way of concealing that quest.

8 *Idea of the Good*, 166.

9 *RAS*, 131. Recall also Gadamer's statement about rhetoric, quoted in the last chapter: "In both rhetoric and hermeneutics, then, theory is subsequent to that out of which it is abstracted—i.e., to *praxis*." ("On the Scope and Function," 80.)

he also speaks of Kant's distinction between "determinative judgment, which subsumes the particular under a given universal, and reflective judgment, which seeks a universal concept for a given particular." He goes on to concur with Hegel's view, that "judgment is really always both. The universal under which the particular is subsumed continues to determine itself through the particular."[10]

Thus both practices—application and articulation—are implied by Gadamer's significant statement that, as the science of the good, practical philosophy "must arise from practice itself, and with all the typical generalizations that it brings to explicit consciousness, be related back to practice."[11] Thus, from the hermeneutical perspective, there is a reciprocity between application and articulation, where these constitute the *events* by which the reciprocity of theory and practice occurs and becomes evident.

By tracing the reciprocity of theory and practice through such historical incarnations, we can now further survey the nature of practice and its relationship to practical philosophy. *Application* brings universals to bear on particulars in such a way that, as Gadamer puts it, we concretize good. Universals and particulars are also related through *articulation*, that is, an event in which one speaks as a witness to universals that have been interpretively inferred from particulars. Articulation does not mean that universals are fantasized truths or mere projections of artificial meaning onto some particular; neither does application mean that good practice is merely a technical imitation of some transcendent, absolute universal in historical form. Rather, we are identifying the events in

---

10   "Afterword," in *TM*, 2nd ed., 557.

11   *RAS*, 92. A similar point is made here, with noteworthy reference to the societal location of the capacity to theorize:

> It is absolutely clear to Aristotle that *theoria* does not have the function of stabilizing *praxis*, but that *praxis*, rather, and practical reason have the function of investing *theoria* with the stability it requires to seek the highest principles—those that justify all other knowledge. It is not my intention to emphasize this as a mere historical reminder, but as a criticism. As long as the liberation of theory is not recognized as a practical and social achievement, we are largely unable to understand how "science" can be mistaken for a method of production, and why it is that practical reason has the power to put theory to a technological use—which, as all will agree, it does, at least in the form of practical unreason. (Hans-Georg Gadamer, "History of Science and Practical Philosophy" [originally 1973], trans. David J. Marshall, Jr., *Contemporary German Philosophy* 3 [1983]: 311.)

which the finitude and contingency of human knowledge participates in the unconditionality of good: articulation does so by bearing witness to universals, and application by concretizing the universals in practice. We can view application as a 'descent' from universals to concrete particulars, and articulation as an 'ascent' from particulars to universals. Neither application nor articulation is more essential than the other and each always 'follows' on the other. While the overt or explicit significance of a given event may be best characterized by one of these terms, the other term can still be a valid characterization of the latent dimension of the event.

The nature of each can be further characterized by other differences. Application pertains most evidently to "what I do in action," and includes all kinds of social practices in which something general, universal or commonly shared is brought to life in the course of interactions among individuals, groups, societies, polities and the natural environment. Articulation appears especially in "what I say in speech," and includes all communicative acts of speech, dialogue, rhetoric, writing, and theorizing in which something is being expressed about or inferred from particular events, and made into something general, universal or shared in common. Of course, to speak is also an act which changes the environment, and to act is also to communicate some meaning to others; the distinction between speaking and acting here is partly metaphorical for the purpose of distinguishing between application and articulation.

These characterizations define practical philosophy as an articulation, in theoretical terms, of descriptive universals *about* practice, and of prescriptive universals *for* practice. But just as it is an articulation which always has roots in experiences of practical application, so it has bearing on prospective practical judgments regardless of how we consciously intend to apply it (or intend not to apply it); practical philosophy has an 'effective history' and, one might say, an 'effective future.' The reciprocity of practical philosophy with practice is always mediated by particular traditions, a mediation which can and should be brought to awareness, insofar as finite consciousness can do so. The practical experiences in which practical philosophy are rooted are made intelligible and meaningful through particular traditions, and these traditional sources are absorbed into the content of practical philosophy; similarly, practical philosophy will be relevant to prospective practical situations to the degree it explicitly brings particular traditions to bear on those situations. Thus, there is no part of practice which is outside the bounds of potential reflection by practical philosophy, and practical philosophy should be involved in such reflection wherever this helps us discern good.

Phronesis is critically involved in the articulation of practical philosophy as well as in its application to practice. It is present when articulation begins, as we clarify to ourselves and bear witness to others about our practical experiences, and when it culminates, as we rationally order the meaning of those experiences in terms of universals. *Phronesis* is also present when application begins, as we face a new practical problem and call upon those universals in order to help target the good in this new situation; application culminates in concretization, that is, in the 'disappearance' of those targeting norms into the concrete good which is appropriate to the situation at hand. The preparation for and results of this practical judgment in turn become a new memory to which *phronesis* once again witnesses, as articulation begins all over again.[12] From this perspective, *phronesis* is not only an Aristotelian virtue but, in effect, a metaphor for the deciding self. *Phronesis* disciplines practical philosophy; yet in making practical judgments it is also *aided* by practical philosophy.

Our primary concern here is to better understand how practical philosophy aids *phronesis* in light of the reciprocity of theory and practice. We propose that practical philosophy renders this aid to *phronesis* through *two basic functions*, each of which involves both articulation and application.

First, practical philosophy functions to illuminate the distinction and the reciprocity between itself as a theory and *phronesis* as a virtue employed in practice. By clarifying this distinction and this dialectic, practical philosophy protects *phronesis* and its authority in practice from all tendencies to transfer that authority to some other faculty or claim, in-

---

[12]   The perspective being outlined in the above paragraphs has some affinity with, and is indirectly indebted to, the framework Paul Ricoeur develops in his essay, "Ethics and Culture: Habermas and Gadamer in Dialogue" (1973), in his *Political and Social Essays*, 243-270 (ed. David Stewart and Joseph Bien, Athens, OH: Ohio University Press, 1974.). Where we have spoken of a reciprocity of articulation and application, Ricoeur's framework concerns the mediation in practical life between imagination and remembrance, between the projects of creativity and freedom and the continuity of tradition and history. We share Ricoeur's view that these contrasting concepts do not indicate antinomies or polarities but mediated realities which imply and require each other. "There are no other paths for carrying out our interest in emancipation than by incarnating it within cultural acquisitions. Freedom only posits itself by transvaluating what has already been evaluated." (269) To put it in positive terms, "The ethical life is a perpetual transaction between the project of freedom and its ethical situation outlined by the given world of institutions." (269)

cluding the claim that practical philosophy itself can better do what is the task of *phronesis* to do. We will refer to this first function of practical philosophy as its *reflective function*, since here practical philosophy acts as the mirror by which to show us, in the language of reason, the nature of practice and the nature of the relationship between practical philosophy and *phronesis*.

Second, practical philosophy functions to assist and guide good practice, and does so by articulating and hypothetically applying universals—such as norms, duties, values and ends—and by articulating the source or ground of such universals. These articulations and applications help direct the attention of *phronesis* to the 'target' of good in concrete situations. This guidance may aid *phronesis* to discern the meaning of existing particulars, and help it to decide what particulars to bring into being, that is, to decide which particular ends and means will concretize the good. We will refer to this second function of practical philosophy as its *participatory function*, since here practical philosophy is serving *phronesis* in its deliberation toward wise practical judgment and thus participating in those judgments. Here practical philosophy is itself actively participating within the scene which practical philosophy mirrors in its reflective function.

This characterization of the functions of practical philosophy should meet with the recognition and approval of Gadamer. They incorporate the principles of reciprocity noted earlier, and, even more explicitly, they closely parallel the "reflective" and "determinative" forms of judgment which we earlier saw him note with approval in Kant, and about which, concurring with Hegel, he asserted that they cannot be separated.

To elaborate on the functions we are proposing, we may speak of the reflective and participatory *contents* of a practical philosophy as those articulations and applications by which a practical philosophy seeks to fulfill its two functions. These contents are not 'neutral' in the manner of modern scientific knowledge, but are sought and asserted for the sake of both understanding and doing morally good practice. Practical philosophy is a science, but in the classical sense, and, as Gadamer states, it "does not allow for the ideal of the non-participatory observer but endeavors instead to bring to our reflective awareness the communality that binds everyone together."[13]

The reflective content of practical philosophy can be further distinguished in its form by contrasting it with the participatory content. The

---

[13]   *RAS*, 135.

reflective content is more a descriptive kind of knowledge than is the participatory content of practical philosophy. The reflective content is more 'reflection *on* practical deliberation' than 'reflection *as* deliberation itself'. It is more related to the preparation of practical deliberation ('what is good practice?') than to the particular contents of deliberation ('what is the practical good in the matter before me?'). So although the reflective content is vitally related to practice, it is related more distantly to practice than is the participatory content, which ultimately blends into action shaped or aided by practical philosophy. And although the reciprocity of articulation and application occurs in the content of each function, the interaction between articulated goals and their application (hypothetical or actual) is more evident in the participatory function than in the more abstract and formal concerns of the reflective function. Consequently, the participatory function of practical philosophy especially proves to be a kind of microcosm of the reciprocity of theoretical universals and concrete particulars which characterize the whole realm of practice. Such theoretical deliberation may assist us in concrete practical deliberation, but can also be selectively appropriated, or modified, or even rejected.

These differences should not obscure the fact that both of these functions serve the purpose of practical philosophy—to help target what is good. It is a fact of life that practical agents both reflect on the meaning of good practice and deliberate on the good at which particular practices could or should aim; practical philosophy addresses these activities through its reflective and participatory functions, respectively. Moreover, each function provides a corrective to the excesses of the other. On the one hand, the participatory function by itself tends to guide us toward the target of practice with too heavy a hand, and can make practical philosophy into something that dictates practice and usurps the role of *phronesis*; the reflective function can help prevent this by reminding us that theorizing cannot provide a technical substitute for practical judgment. On the other hand, the reflective function by itself will tell us about the formal or abstract character of good practice but can also tend to become distant from, and ultimately irrelevant to, concrete issues; the participatory function can help curb this tendency by focussing attention on the concrete problems which require our decision.[14]

---

[14] This benefit of the participatory function, and its justification by the reciprocity principle, are reflected in the following interpretation of Gadamer by Antonio Da Re, in his 1982 work (also quoted in Chapter Four, note 104), *L'ermeneutica di Gadamer e la Filosofia Pratica* (*The Hermeneutics of Gadamer and Practical Philosophy*) (Rimini, Italy:

It seems reasonable, in sum, that by addressing and fulfilling these two functions, practical philosophy will affirm the hermeneutic principle of the reciprocity between theory and practice, and so will best fulfill its purpose. To recapitulate, practical philosophy involves, first, *reflection on* the reciprocity of theory and practice, and on its significance for both theorizing and for practical action. The reciprocity of theory and practice indicates, second, that practical philosophy should be directed to *participate in* this reciprocity, since we know that its theoretical assertions are inseparable from either preceding memories or prospective tasks. By these means, practical philosophy helps to keep open the space where concretization is to occur, and steer us toward the best concretization within that space. This should happen, of course, not only in practice proper, but also in the practice of thinking about practice—in which we are currently engaged.

It is helpful to use metaphorical images to understand the task of the practical agent, and the role which practical philosophy can play in deliberation. We have seen Gadamer use Aristotle's metaphor of the archer. A rather different metaphor may also be useful, that of the architect. The architect plans and then builds 'good space' as we use practical philosophy and then *phronesis*: in each act of creation, the architect uses prior

---

Maggioli, 1982). The passage also happens to corroborate the view developed in the preceding chapter concerning the defense of Gadamer's position vis-a-vis Habermas' critique:

> One could object to Gadamer, and indirectly to Aristotle, that their attempt to adhere to human conditionality by putting aside abstract and unattainable normative theories leads them to the serious danger of legitimating the status quo, even when injustices and abuses occur. But it seems to me that Gadamer tries to bypass this obstacle by replying that Aristotle's practical philosophy is not simply a method, a technique or a chart of values, but in a sense is actually theory itself, and therefore also has critical capacity: it therefore has nothing of the modern conception of theory, which is understood from a totally utilitarian perspective as pure and simple application to practice, and is judged solely on its visible results, but it is rather a practical philosophy which, by dipping into practice, reclarifies practice itself, within the rational critique of theory, so that later it can relate itself back to practice. Essentially, then, the contrasting and unending relationship of theory and practice seems to give the right significance to the description of experience as a whole, grasped without illegitimate deformation, and leaving open the way that leads to what is strictly prescribed by what 'must be' [*dover essere*] in human existence. (113-114, translated by Cristina Puglisi Kamitsuka.)

training and experience to concretize, as it were, one particular good space; in planning and in execution, however, no two structures are alike; each serves different needs in a different environment, and each is subject to different pressures and limitations. The architect must deal with all of these factors and still seek, as far as possible, to make the structure a thing of beauty and a home to some human activity. Similarly, in practice we must deal with changing situations, with limitations and obligations, as we aim to 'create' practical good, another kind of human home; our education in practical philosophy should help orient and guide us in any of these situations, but cannot tell us how to decide specific or new problems. And while some persons are architectural authorities, everyone can recognize both awesome and abysmal structures; likewise, while some study ethics as a vocation, everyone has ethical sense and capacity.

ASSESSMENT OF GADAMER'S PRACTICAL PHILOSOPHY: FOUR DEMONSTRATIONS

If we look at Gadamer's work from the perspective of the preceding account of the functions of practical philosophy, what do we find? Most of Gadamer's writings we have examined up to this point fall under the reflective function of practical philosophy. Indeed, in outlining our concept of practical philosophy's two functions, we have accepted a basic if not central message of Gadamer's reflective content: practice is different from making, and theoretical efforts to help us target the good must take this into account. Similarly, the activity of *praxis* is not synonymous with the activity of *poiesis* (making), just as *phronesis* is distinguishable from *techne*: choices which concretize good are fundamentally different from making products by technically replicating a design. Since only *phronesis* and not *techne* can make these decisions, a corresponding restraint of our expectations of theorizing is required, lest we allow theory to usurp the place of *phronesis* and technicize practice. Consequently, Gadamer strongly emphasizes the necessity of interpretation and judgment in application. As Gadamer puts it in negative form, there are no rules for applying rules, meaning that there are no technical criteria which can ensure that general rules about practical good will be correctly applied.

With regard to the participatory function, it is apparent, at least in the materials examined so far, that Gadamer pays little attention as compared to his attention to the reflective function. The participatory content he does present is usually expressed in very general and very brief terms, and always has a significance secondary to that of the reflective function. The constructive application of norms and values which do appear remain nebulous. Moreover, while we have seen that he seems confident that the realization of solidarity is always possible, the basis for this confidence remains obscure.[15] These are matters which a practical philosophy must address through its participatory function if it is to be regarded as complete or at least a coherent whole. Our inquiry so far, then, poses the preliminary judgment that Gadamer's practical philosophy is imbalanced, that it is disproportionately developed in favor of the reflective function.

Is there, then, more to Gadamer's practical philosophy that we are missing? Are there other texts of Gadamer's where we might find the participatory function addressed more substantively and directly, or at least addressed in some way? Are there any other texts at all that deal with practical philosophy, or bear on its purposes and functions? More specifically, does he tell us more about what he means by solidarity, and if not, why not? Or can he tell us more about the basis of solidarity, about his hope for its possible realization in human practice? In regard to any of the practical problems facing us today, how would he assess its nature, cause and correction? And in general, how would he apply his practical philosophy?

All of these questions—and the fact that such questions must be raised—certainly invite further analysis of what Gadamer does say, and of what he does not say. The rest of this chapter focusses on texts and statements by Gadamer which are new to this inquiry and can help us to answer these questions—or at least to know if further inquiry of Gadamer is fruitless. Our concern here, then, is with what Gadamer does say, beyond what we have already surveyed, and with what, if anything, it further reveals or further substantiates about the practical philosophy he seems to have in mind. The analyses contained in the following pages concern four topics: first, Gadamer's critique of the modern deformation of theory and practice; second, reflections on the problems of global order and disorder in the modern age; third, the nature of the confidence Gadamer thinks we can have in the possibility of solidarity; and fourth,

---

15 We may also note that, even to understand the relationship between solidarity and the intellectual virtues, it was necessary to rely on the work of Ronald Beiner, although it might be expected that Gadamer would be particularly concerned to develop this primary link.

the relationship between the conditional and the unconditional in theorizing which is done to aid moral deliberation.

The Modern Deformation of Theory and Practice

Among Gadamer's more recent works which bear on practical philosophy, one of the more frequent themes has been a comparatively impassioned critique of our modern practices, and in particular of how we conceive practice and theory, and how these distorted conceptions have damaged our whole social life. Gadamer clearly thinks that this deformation—in our thinking and in our doing—is a source, if not the primary root, of many of our modern practical problems. It thus constitutes the major critical (as opposed to constructive) focus of his ethical reflections. However, Gadamer has not translated this concern into any systematic treatment of the issue. His statements range from the analytical to the polemical, from the general, even the over-generalized, to the anecdotal, and all are scattered throughout a number of essays. What follows is an attempt to synthesize and order his thoughts on this topic.

Gadamer writes that today, theory is synonymous with science in its modern sense, that is, "a knowledge of manipulable relationships by means of isolating experimentation."[16] Practice is understood as the "application of science to technical tasks"[17] and is "defined by a kind of opposition to theory."[18] Thus, the modern view polarizes theory and practice and makes theory the primary authority for the determination of practice.

According to Gadamer, this polarization of theory and practice is the culmination of changes in their original meanings which began in the seventeenth century, with Descartes and Galileo. The Cartesian legacy is that experience ceased to be the presupposition of science and instead became its object, while Galileo's abstraction of causes from what is observed in experience leads to a technical perspective and away from the classical sense of science as an encompassing knowledge of reality.[19] These foundations of the modern notion of science eventually had revolutionary consequences for what is meant by 'theory' and 'practice.' The result in our age is that while science "renounces comprehensive knowledge in the grand Aristotelian style," and thereby loses the ability to

---

16   *RAS*, 70.
17   "Hermeneutics and Social Science," 312.
18   *RAS*, 69.
19   "Theory, Technology, Practice," 533-534.

provide a "whole orientation to the world,"[20] it gains "control over nature in an entirely new dimension and an entirely new sense."[21] The creation by science of notions of theory and practice which supplanted their classical meaning constitutes, therefore, "a true event in the history of man, which conferred a new social and political accent upon science."[22]

Two grounds for objection to these modern concepts of theory and practice are evident in Gadamer's writing. The first is internal to the logic of modern science and the second concerns society's relation to this science. Concerning the first point, the historical development of modern science makes clear that "this science was primarily a knowledge of the possibilities of controlling natural processes."[23] It rested on the belief that "the possibilities for truth in science were determined by method." In Gadamer's view, the use of method is not to be proscribed, even in the social sciences, but carries with it a special responsibility: "Every scientist who gives an account of the conditions of his own procedure knows the price he pays in return for the certainty, controllability and solid advance of his investigations."[24] But this responsibility to identify the costs of scientific methodology is eroded by the philosophical presupposition that methodology is applicable to any object.[25] The human pursuit of knowledge thus becomes slave to a methodological logic that interprets everything according to its own criteria, and as this has occurred, science "of its own accord [has] passed over into the limitlessly widening regions of human *praxis*."[26] And "the unavoidable consequence is that science is invoked far beyond the limits of its real competence."[27]

---

20 *RAS*, 144.
21 "The Power of Reason," 8.
22 "Theory, Technology, Practice," 534.
23 "The Power of Reason," 9.
24 "Hermeneutics and Social Science," 310.
25 Modern science, says Gadamer,

> raises the claim that on the basis of its methodological procedure it is the only certain experience, hence the only mode of knowing in which each and every experience is rendered truly legitimate. What we know from practical experience and the 'extrascientific' must not only be subjected to scientific verification but also, should it hold its ground against this demand, belongs by this very act to the domain of research for science. There is in principle nothing which could not be subordinated in this manner to the competence of science. ("Theory, Technology, Practice," 530.)

26 "The Power of Reason," 60.
27 "Theory, Technology, Practice," 548.

Modernity is thus marked by an intellectual idolatry of science, of which one effect is the distortion of the concept of practice itself.

The social and political forms of this idolatry of science constitute the second general cause of the predicament of modern practice. The disintegration in our century of common traditions of value has "fostered a new desire and inner longing in our society to find in science a substitute for lost orientations—a very dangerous situation."[28] The successes of science in controlling nature "engenders the mounting expectation that science is ultimately capable of banishing all unpredictability from the life most proper to society by subjecting all spheres of human living to scientific control." Ours is an "age of new, radicalized faith in science," which seems to ignore the failures of science to meet such expectations. But finally this faith in science signifies a human desire to be relieved of the responsibility to act and to decide,[29] and instead "invests the expert with an exaggerated authority."[30]

Both the intellectual and social forms of the idolatry of science contribute to the atrophy of true practice. While we are taught to think that science has made practice more effective and human good more secure, Gadamer claims it is quite often the opposite. In thinking that the application of science fulfills the function of practice, modern society has changed the meaning of practice. "In all the debates of the last century practice was understood as application of science to technical tasks. That is a very inadequate notion. It degrades reason to technical control."[31] Of course modern science's intended focus is not on moral practice but on natural causality, and in the sense that knowledge of causality introduces the possibility of controlling and directing causal forces toward new effects, science "means not so much knowledge as know-how."[32] But it is easy to confuse this with practice since know-how has value only in application. In fact this false identification of scientific know-how with moral practice is, for Gadamer, a central feature of modern technology:

> It would appear to me . . . that science makes possible knowledge directed to the power of making, a knowing mastery of nature. This is technology. And this is precisely what practice is not. For . . . [technology] is not knowledge which, as steadily increasing experience, is acquired from practice, the life situation, and the circumstance of action. On the contrary, it is a kind of knowledge which for the first time makes possible a novel relation to prac-

---

28 "Hermeneutics and Social Science," 307.
29 *RAS*, 147, 148.
30 "Hermeneutics and Social Science," 312.
31 "Hermeneutics and Social Science," 312. See also "Afterword," *TM*, 2nd ed., 556.

tice, namely, that of the constructed application. It is the essence of its procedure to achieve in all spheres the abstraction which isolates individual causal relationships.[33]

The power and success of technology make it a natural candidate for the focus of a new concept of political order: technocracy. "A novel expectation has become pervasive in our awareness: whether a more rationalized organization of society or, briefly, a mastery of society by . . . more rational social relationships may not be brought about by intentional planning."[34] But, as Gadamer clearly connotes, this technocratic ideal has an ideological character: "The ideal of the full mastery of the tasks and problems of our civilization by science conceals an insoluble contradiction between the role and function of the expert . . . and the fact of his own membership in society."[35] The public expects the expert to be a "substitute for practical and political experience," and permits the expert to make decisions on its behalf. In other words, we are taught to accept the authority of the experts, and thus look to them to fulfill their promises. However, this is an expectation which, "in light of a sober and methodical self-appraisal and an honest heightening of awareness, [the expert] cannot fulfill."[36]

Thus the technocratic features of society stifle the reasoning of all its citizens, not just of its managing 'high priests.' Gadamer's deepest concern is with the ways in which science and technology destroy public discourse, the vital link between reasoning and good social practice. In effect, what is being destroyed is the public space for dialogue, and therefore for friendship and solidarity as expressions of 'judging-with.'

In a scientific culture such as ours the fields of *techne* and art are much more expanded. Thus the fields of mastering means to pre-given ends have been rendered even more monological and controllable. The crucial change is that practical wisdom can no longer be promoted by personal contact and the mutual exchange of views among the citizens . . . Immediate and natural interaction in the course of daily life is no longer the unique source and dominant mode for the elaboration of common convictions and normative ideas.[37]

---

32 "Theory, Technology, Practice," 534.
33 "Theory, Technology, Practice," 534.
34 *RAS*, 72. See also "Power of Reason," 13.
35 "Hermeneutics and Social Science," 310.
36 *RAS*, 72.
37 "Hermeneutics and Social Science," 313-314.

The mediation of information within public life increasingly indicates "the technologizing of the formation of public opinion. Today this is perhaps the strongest new factor in the play of social forces." Those who select, from the growing mountains of available information, what is presented to the public hold a great power, and "can intentionally steer public opinion in certain directions and exercise influence on behalf of certain decisions."[38]

Gadamer's analyses of the manifestations and consequences of modern science and technology are summed up in an ironic "general rule": "The more rationally the organizational forms of life are shaped, the less is rational judgment exercised and trained among individuals."[39] In other words, the more we rely on 'constructed application' derived from the logic of modern science, the less we are even aware of the reasoning of *phronesis* which conceives of application as concretization.[40]

In several essays about modern practice, Gadamer gives particular attention to the question of the responsibility of scientists. This is also a topic which gives him an opportunity to propose constructive remedies for the state of modern practice. And since he often criticizes the technocratic tendencies of scientists, it is reasonable to hope Gadamer might provide some further indication of how modern scientists of the natural and social worlds should conceive of their social roles and responsibilities. He points out that although the notion of the responsibility of scientists has become popular in recent years, it is "fundamentally nothing new." Classical reflection on the extent of this responsibility "sought

---

[38] *RAS*, 73. We can expand on Gadamer's statements about contemporary social communication: the problem is not merely that the content of public discourse is prejudiced toward the technocratic ideal, but that the rhetorical forms for pursuing discourse are already manipulated by technical reasoning. The very forms of communication on which a clear understanding of practice depend are now themselves often carriers of the technocratic prejudice which distorts that understanding.

[39] "Theory, Technology, Practice," 546.

[40] Gadamer elaborates on this:

> While the development of practical technologies, to judge from appearances, diminishes the distance between the general knowledge of the science and the correct decisions of the moment, the qualitative difference between practical knowledge and scientific knowledge actually gets larger. Precisely because the technologies which are applied are indispensable, the sphere of judgment and experience, out of which the right practical decisions are made, gets smaller. ("Theory, Technology, Practice," 550-551. Also see a similar statement in Hans-Georg Gadamer, "Notes on Planning for the Future," *Daedalus* 95 (Spring 1966): 582.)

its answer in the domain of 'practical philosophy' by subjecting all the 'arts' to 'political' ordering." Thus if the scientific specialist "becomes mindful of the limits of his specialization," he can become "ready to acknowledge experiences which are uncomfortable for the private interests of the researcher—such as, for example, the social and political responsibility present in every profession where others are dependent upon someone." Today, "there is need of this on a worldwide scale," particularly because "social-political consciousness has not kept pace" with the developments of the technical world.[41]

This is merely a statement of the general problem and of the general solution. Is Gadamer able to further specify either of these? Here his position becomes ambiguous. On the one hand, he criticizes the philosophy and methodology of modern science for its presumption that all non-scientific knowledge is of no value unless it can be converted into 'scientific' form, a conversion which, in Gadamer's eyes, necessarily deforms and conceals the knowledge of the limitations of science.[42] He also holds those who mediate scientific knowledge to socio-political uses particularly responsible for the deformation of modern practice.

On the other hand, Gadamer defends the legitimate scientific work of the "serious scientist" from the charge that he or she is guilty of perpetuating a technocratic ideology, since a correct understanding of science includes knowledge of its limitations.[43] In an oral discussion of his essay, "Hermeneutics and Social Science," Gadamer has this to say: "I find it absolutely unjust to accuse scientists of the abuse of science. They cannot as scientists find anything in the sciences which limit or negate the possible abuse of the results."[44]

Of course, they cannot—not as scientists; but this is not the problematic issue. It is fine to illustrate these two approaches to judging the work of scientists, but what should be done when these two assessments collide? Gadamer gives little attention to the more difficult task of a single perspective in which to correctly understand both the abuse of science and the legitimate use of science. Unfortunately typical of Gadamer's efforts at resolution is the following statement, which only identifies the issues once more, but does not address them: "One can demand of the scientist insofar as he is a citizen that, as a citizen, he act in the public interest; however, insofar as he is a scientist this demand is

---

41 "Theory, Technology, Practice," 553, 554.

42 "Theory, Technology, Practice," 530.

43 "Hermeneutics and Social Science," 307.

44 Hans-Georg Gadamer, "Summation," *Cultural Hermeneutics* 2 (1975): 329-320.

inapplicable.[45] Gadamer's general aim seems to be both to protect science as well as to chastise it, and the quotation illustrates his indecision about how to do both. It is quite true, as Gadamer would no doubt reply, that *phronesis* must be relied on to decide how to resolve situations in which, for example, one's responsibilities as scientist and as citizen conflict. But between the generality of Gadamer's pronouncements and the particularity of *phronesis*, there is a wide spectrum of opportunities for targeting what is good, a targeting which would benefit *phronesis*.[46]

Gadamer recovers what seem, at least to us, to be his own best instincts when he says that "the abuse of the sciences, in my view, occurs at the level of political reason,"[47] and therefore not fundamentally at the level of scientific reason. As he says elsewhere, "It is not the task of science but of politics to supervise the application of the know-how made possible by science."[48]

When we synthesize these and other statements, it seems Gadamer is arguing that scientists alone should be responsible for pure research, while decisions about applied research and the social use of that scientific knowledge should finally be a political matter and therefore in the hands of some political authority, as yet unspecified. This arrangement has a certain reasonableness to it, but it is still far from having concrete practical utility. *Whose* political reason is responsible for the abuse of science and for its correction? And *in what form* is politics to supervise the use of science but not its research? The distinction between pure and applied research is notoriously unclear, allowing all sorts of economic, political and national interests to penetrate quite deeply into scientific discovery. Whose responsibility is it then?

Obviously, Gadamer has not posed such concrete questions in his own work. And it is only fair to note that several of our quotations come from spontaneous oral remarks. These considerations may excuse some of his generality and ambiguity. But why has he not given the matter more attention? The topic is one he himself raises, and it lies squarely within the jurisdiction of practical philosophy. And if he thinks nothing

---

45 "Summation," 329-330.

46 See also Gadamer's short essay, "Science and the Public," as an illustration of a collection of possibly meaningful insights into this topic which never quite coheres as a whole nor identifies any clear questions for further analysis, much less practical judgment (*Universitas*, 23, no. 3 (1981): 161-168.)

47 "Summation," 329.

48 "Theory, Technology, Practice," 555. He has also written: "The real responsibility of science . . . cannot be met by any science as such," which is ultimately an appeal to *phronesis*. ("Power of Reason," 12.)

more can be said on the level of practical philosophy, he does not indicate this either.[49]

Global Order and Disorder

We turn now to a single text, a 1965 address entitled, "Notes on Planning for the Future," in which Gadamer has what seems a particularly good opportunity to expand on his practical philosophy, and on how it should fulfill the participatory function of such a philosophy.[50]

While Gadamer's debate with Habermas seems to have stirred Gadamer to give more attention to contemporary practical issues, this essay shows that many of his basic themes regarding practical philosophy and current social issues predate that debate. No doubt the conference to which Gadamer was contributing had something to do with his choice of topics and the way he addressed them. This was a week-long conference on the "Conditions of World Order," held in 1965 in Bellagio, Italy, and attended by some two dozen participants—mostly academics, but several with governmental experience as well. [51] Here was certainly an oc-

---

[49]   In the absence of other sources, one small observation from the hermeneutical perspective may be appended to this topic. It is true that the primary intent of certain kinds of scientific research has no immediate or intended social bearing; yet doing research is already a human practice as evidenced by the presence of hermeneutical judgment (see: "Theory, Technology, Practice," 558, 560; "Power of Reason," 12, 14.) Therefore, engaging in scientific research is a product of human choice and in choosing it one necessarily exercises moral judgment; and if one continues to select a particular course of scientific research, one must do so with moral wisdom. We can choose whether or not to do science; contrary to some opinions, it does not choose to do us.

There is indeed a religious or divine dimension to science, and to the wonder and inquiry which impels it; but we may want to reconsider whether this awe necessarily entails the imperative to discover everything or anything. The scale of costs and the scope of choices which science will bequeath to our descendants in Malthusian proportions are other, and by no means insignificant, reasons of utility to consider in making our decision: how much science, or which science, is good science?

[50]   *Daedalus* 95 (Spring 1966): 572-589. The title seems to hint at the technocratic notion of predetermining the future, beyond the idea of simply preparing for the future. (The translator, if any, is unidentified. The original publication, as well as original language, of the essay is unclear. A German version, under the title, "Über die Planung der Zukunft," appeared the following year—in 1967—in Hans-Georg Gadamer, *Kleine Schriften*, vol. 1 [Tübingen: J. C. B. Mohr (Paul Siebeck), 1967], 161-178, where, to the reader's confusion, a footnote refers to the earlier *Daedalus* text as a translation.)

[51]   Notable names among them were Stanley Hoffman, Henry Kissinger, Jan Tinbergen and Raymond Aron, who served as the moderator. They came primarily from the fields of political science and economics, with a few other fields also repre-

casion to encounter and converse with intellectuals who, to varying de-
grees, may have represented the very technocratic inclinations we have
just seen Gadamer criticize. In his address, Gadamer presents a critique
of the technocratic perspective on questions of global order and disorder,
and attempts to bring his practical philosophy to bear on them. A number of the main points are worth mentioning here.

Gadamer develops his theme that a basic contemporary problem is a
concept of practice which is distorted by the technological model of practice as making a product. He then offers the model of the pilot as an
alternative. He asserts that this model accounts for two central features of
practical judgment. First, in every situation practice has a certain limited
range of alternatives open to it and must reckon with the forces which
define that range. In Gadamer's words, practice requires the "main-
tenance of equilibrium, which oscillates in a precisely set amplitude."[52]
Of course, we might add, for the model based on 'making' (*techne*) there
is also an amplitude of possible practice. The implied point, however,
can be expressed in terms of understanding 'making' as an activity that is
implosive, that reorders and uses the environment with less regard for
the surrounding consequences than for creating a certain product.
Piloting is not like this in that the choice of a craft's course can never be
abstracted from consideration of its environment but must find the best
way to 'ride with it' in order to 'ride through it.' If piloting were instead
undertaken as an implosive activity it would leave the craft vulnerable to
destruction by environmental forces. This leads to the second feature of
practical judgment which, in Gadamer's view, is addressed by the model
of piloting. It accounts for what practice aims at, that is, for "the selection
of a direction which is possible within the oscillating equilibrium."[53]
Gadamer seems to be restating in different terms the idea that the
dialectic of logos and ethos must be respected if practice is to be genuine.

The critical issue of practice in our age, Gadamer asserts, is to 'pilot'
our way through the tension between scientific know-how and practical
wisdom. A number of professions necessarily involve dealing with this
tension, such as law, psychiatry, and teaching, but medicine best epito-
mizes the contemporary challenge of practice. The physician's "human

---

sented, among whom Gadamer was the sole philosopher. This and more information
about the results of this fascinating conference can be found in: Stanley Hoffman,
"Report on the Conference on Conditions of World Order," *Daedalus* 95 (Spring 1966):
455-478.

[52] "Notes on Planning for the Future," 582.
[53] "Notes on Planning for the Future," 582.

function must not be submerged by the merely 'scientific,'" and he or she "must strike a balance between practical activity and science." For "here modern science, in all its magnitude and promise, collides directly with the historic role of the healer and the assumption underlying his compassionate activity." While other forms of pragmatic knowledge can be verified directly in the results of action, "the task of medicine remains irreducibly ambiguous" because the physician never knows for certain which aspect of the results were due to intervention and which to nature assisting itself. This is exemplified in the unavoidable vagueness of the concept of health, which since ancient times could only be characterized as a kind of balance or equilibrium. Medical intervention, he implies, should be considered only as the attempt to aid in reintroducing that equilibrium when it has been disturbed, thus putting "a special set of limitations" on intervention.[54]

In other passages, Gadamer raises social and political issues about which the world clearly needs wisdom for the piloting of our practices. He observes that the widely shared project of a world order seems to be based on the belief that this goal can be reached by gradually accumulating successes of global uniformity in smaller discrete areas of life, such as health or communications. But, he asks, "does talk of creating world order still make sense if, from the start, we are faced with irreconcilable ideas on the constitution of a right order? . . . Does not all planning on a world scale depend on the existence of a definite mutual conception of the goal?"[55] In fact, there is no such universal agreement about the goal. It is illusory to assert, for example, that the prevention of "global self-destruction" or the elimination of "economic disorder" is the goal of world politics, for this begs the question of what peace is or what economic order is, and thus are empty standards for political judgment. Peace, especially in the nuclear age, comes to mean merely that "the *status quo* is the international order to be preserved . . . But is that a meaningful standard for politics?" As to economic order, "we still have not resolved" the conflict between capitalist and communist prescriptions for it.[56] (Now, of course, this conflict seems much closer to resolution, at least in the West. But it did not happen by means of the kind of 'rational'—that is, technical—negotiation and planning which was no doubt anticipated by many in 1965—or in all the years following, until recent events surprised everyone. But this reversal of expectation is perhaps implied in what

---

54 "Notes on Planning for the Future," 583, 584.
55 "Notes on Planning for the Future," 572-573.
56 "Notes on Planning for the Future," 574, 573.

Gadamer said, if not intended: if such a means has not resolved this conflict by now, there is probably a deeper reason, and a deeper problem which is being ignored or avoided.)

Such a critique of the hopes and plans of the technocrats of the world evokes the expectation that Gadamer will propose some definitions of world order. However, Gadamer in effect challenges the technocratic assumption that such definitions can be operationalized and made the guaranteed means of practice:

> Is it possible to think of one precise political arrangement which would not provoke all the very opposite of ideas of world order? Are political visions of order conceivable which do not favor one political system and [do] that only when it is detrimental to the other? Should one assert that the existence of contradictions among power interests constitutes disorder? Is not such a condition itself constitutive of the essence of political order?[57]

The experts of the world may believe that, if given the chance, their technical rationality could solve many of the world's problems. But for Gadamer the remedies they would seek are so preconceived, or so foreign to, or so little informed about the practical realities 'on the ground' in places far from their own, that the reaction to and results of such efforts will generate new conflicts and disorder. The failures of global politics, such as the elusiveness of international agreements, are not, as the technical experts may think, the 'fault' of political authorities. And even if such experts deny it, there are few issues indeed which can be resolved on purely technical grounds.[58] He suggests that the task is incorrectly posed by the experts. We must ask

> whether it is not precisely the overdependence on science in our business and social life . . . that has increased the uncertainty regarding the intended goals, the content of a world order as it should be, by first subjecting the design of our world to scientifically informed and guided planning, while obfuscating the uncertainty which surrounds the standards.[59]

These observations point back to a critical implication at the end of the preceding quotation. Gadamer frames his rejection of the technocratic ideal by transmuting the meaning of 'order.' Ignoring its conventional meaning, he refers to certain of the conflicts of powers within and between societies. What Gadamer seems to mean, then, by 'order' may in

---

57 "Notes on Planning for the Future," 574.
58 "Notes on Planning for the Future," 580.
59 "Notes on Planning for the Future," 575.

fact be widely perceived as a kind of disorder, not of violence necessarily, but of the fluidity and constant transformation that comes with dialogue, challenge and negotiation. This view has further hermeneutical significance for practical philosophy which Gadamer does not elucidate. The seeming 'disorder' among disparate communicating selves contains a truth about an elemental order which unites humans and appears in their linguisticality. Here dialogue appears not just as the occasion for resolving misunderstood or foreign meanings, but a process which incarnates and embraces the unending conflicts between different interests, on both individual and corporate levels. Politics, then, is not the suppression of 'disorder' into a perpetually happy—or in the hermeneutical context, silent—and static order. Rather, politics is a means for providing and legitimizing forms of dialogue by which 'disorderly' differences and conflicts can be lowered or elevated, resolved or celebrated, as seems fitting, so that the social reality will best concretize the order of being. Gadamer thus implies that there are important forms of political disorder which are mandated by the 'order' of practical good, and thereby points toward an understanding of politics which is an alternative to both the technocratic order of the expert and to the orderliness of simply basing practice on "conformity to the established social order and its standards."[60]

This perspective is evident elsewhere in the essay as well, in Gadamer's objection to the ideal of a universal language and in his affirmation of social and cultural diversity. In fact, these points turn out to be his primary articulations in this essay about how we should pilot our way through the modern tensions between science and practical knowledge. If there was any thought that the hermeneutical 'fusion of horizons' meant a safe homogenization, then Gadamer clearly excludes such an interpretation here. To Gadamer the problem inherent in the desire to achieve, for example, a universal artificial language, is the same problem as in any other aspect of the aim of complete global administration: "it might be unavoidable that the universal means would become universal ends." Contrary to some expectations, it would not make possible the communication of "everything imaginable"; rather, it would "ensure that only the data assimilated in the programming would be communicated, or even thought."[61] In contrast to such an artificial language, Gadamer argues, as he does in *Truth and Method*, that natural language contains an "inner infinity" which "perfects and develops itself in association with

---

60 "Notes on Planning for the Future," 581.
61 "Notes on Planning for the Future," 585, 586.

the living traditions that encompass historical humanity." And confirming again our argument that Gadamer's ontology is dynamically present in his analysis of social practice, Gadamer asserts that the possible success in the future of an artificial language might increasingly conceal but will never eradicate the infinite possibilities of the dialectic of natural language.[62] By means of this appreciation of natural language diversity, Gadamer hopes we will increase our awareness of the "differences between peoples and nations," that is, of the concrete diversities among humans which the experts would rather eliminate, a task of learning tolerance that is "now hardly ever performed by science."[63] In this latent way, he again invokes the value of certain kinds of 'disorder' in the world. He speaks of such reflection as a task of intellectuals and philosophers, the results of whose labor "can become the conscious property of everyone."[64]

As in our earlier analysis of Gadamer's critique of modern practice, this essay elicits a mixed reaction. Gadamer's primary attention here is on exposing and correcting the technocratic approach to global issues. This critique is, again, largely an expression of the reflective content of his practical philosophy. In one respect, however, Gadamer does coordinate reflective and participatory claims. The plurality of traditions is presented as an ineradicable given as well as a value, and this clarifies his opening observation on the absence of common goals or concepts about world order: he means to remind us that global good is not something we can technically engineer. Our goal should rather be an awareness of the deeper and larger equilibrium to which certain kinds of 'disorder' actually contribute, and in terms of which our practice will better target the good.

Still, the values he articulates in this address remain quite abstract. The practical meaning of 'disorderly' order, of pluralism, and of tolerance among traditions and languages are things which can be elucidated only by giving the agent—or the pilot, to use Gadamer's metaphor— more concrete navigational rules and more specific 'compass bearings' about the social arrangements and political institutions which would help realize those values. Moreover, practical philosophy must eventually address complex questions about knowing when good disorder becomes bad disorder or vice versa, or about how to deal with conflicts which may arise between the value of plurality and the value of solidar-

---

[62] "Notes on Planning for the Future," 586, 587.

[63] "Notes on Planning for the Future," 587-588.

[64] "Notes on Planning for the Future," 588.

ity (or community). In fact, by neglecting these problems, his affirmation of solidarity harbors the possibility of being at odds with his affirmation of pluralism. It is just such questions which, one hopes, would be of particular concern to those who have political responsibilities. At their best, such figures will look to a philosopher, not for ready-made solutions (which they would neither expect nor tolerate), but for applicable insights which go beyond where Gadamer stops.

Hope for Solidarity

Perhaps our search for further participatory content in Gadamer's practical philosophy will have better success on a different level of inquiry. Does Gadamer ever elaborate further on the moral values which should be among the targets of practical philosophy? There are a few instances in which Gadamer identifies such values, asserting or implying their validity. But other problems often accompany these cases. For instance, in one place he speaks of *phronesis* as being involved with practical decisions about "happiness, health, peace, freedom and other stable factors of human-being-in-nature."[65] Nothing more is said about these values; it seems to be only a list intended to establish some point of reference with his audience (with the exception of the last item, which is perhaps a term of Gadamer's own creation, or at least is not a conventional modern term).

There are also other scattered references to freedom in particular.[66] Some may find more significance in such statements than they really contain. No doubt Gadamer sincerely believes in the value of freedom; but such statements seem rather perfunctory, as though he is not vitally concerned with such matters, or feels obliged to assuage any concern about whether he is in fact a citizen of the west in the twentieth century. This may explain why no attention is paid elsewhere to the concept of freedom, and why no inquiry is made into the conflicts and controversies involved in actually spelling out what freedom means, in theory or in

---

65 "Hermeneutics and Social Science," 313

66 In *RAS*, Gadamer writes that "there is no higher principle than that of freedom . . . And we understand actual history from the perspective of this principle: as the ever-to-be-renewed and the never-ending struggle for this freedom."(9) Elsewhere, he says that he is "in favor of a government and politics that would allow for mutual understanding and the freedom of all." ("Appendix. A Letter," 264.)

practice. If Gadamer feels there is a vital center to the moral life, it must be somewhere else.[67]

And indeed it is: there is more in Gadamer's texts by which to confirm that if any value is elevated by Gadamer, it is the value of solidarity. In Chapter Four, it was inferred from Gadamer's writing that he believes solidarity is not simply the product of social conditions, and instead he seems confident that solidarity is always possible. Gadamer's statements in two additional passages do more than substantiate these inferences if we invest some patience to reveal what Gadamer says but leaves unspoken; they also imply the reasons and ultimate basis of his hope for solidarity. In this respect, the passages at hand are among the few in which Gadamer touches on what in other contexts might be called meta-ethics or philosophy of history.

One of these passages occurs in a letter which Gadamer wrote in response to two of Richard Bernstein's essays about him, essays which Bernstein largely incorporated into his 1983 book, *Beyond Objectivism and Relativism*.[68] Bernstein is sympathetic to a number of Gadamer's themes but has certain objections, not unlike Habermas' to Gadamer's practical philosophy. Of particular relevance here, Bernstein doubts that the traditions on which *phronesis* depends for nurture still have sufficient vitality today. He thus implies that solidarity is not attainable in history simply through reliance on *phronesis*: when those nurturing traditions are absent, so too is *phronesis*.

> Given a community in which there is a living, shared acceptance of ethical principles and norms, then *phronesis* as the mediation of such universals in particular situations makes good sense.
>
> The problem for us today, the chief characteristic of our hermeneutical situation, is that we are in a state of great confusion and uncertainty (some might even say chaos) about what norms or "universals" ought to govern our practical lives. Gadamer realizes—but I do not think he squarely faces the issues that it raises—that we are living in a time when the very condi-

---

67 A footnote by Gadamer in *TM* has some relevance here, and of particular interest as well, because Leo Strauss is quoted. In the 2nd edition, 271, note 189, in support of the rehabilitation of the term 'prejudice,' Gadamer cites with approval Strauss' *Die Religionskritik Spinozas*, 163, and offers this quotation: "The word 'prejudice' is the most suitable expression for the great aim of the Enlightenment, the desire for free, untrammeled verification; the *Vorurteil* [prejudice] is the unambiguous polemical correlate of the very ambiguous word 'freedom.'" This reference to Strauss reinforces the view that freedom, in the form it is bequeathed to us by the Enlightenment, is not Gadamer's highest value; after all, he has sought to rehabilitate what the Enlightenment considered the opposite of freedom.

tions required for the exercise of *phronesis*—the shared acceptance and stability of universal principles and laws—are themselves threatened (or do not exist).[69]

Echoing Habermas, Bernstein argues that if the conditions for *phronesis* do not exist and if practice in its true sense has been so deformed by the technocratic ethos, then we need a social scientific theory which could identify the conditions destructive to *phronesis,* and which could thereby aid restoration of *phronesis* in society today.[70] The larger issue at stake here hinges on whether one thinks the formation of such a theory is possible in the absence of *phronesis* (in which case, one wonders why we need bother over resuscitating this virtue) or whether the creation of such a theory already demonstrates the presence and activity of *phronesis* (in which case, it is not really the theory alone which restores authentic practice and community, but the prior—and apparently uncaused—*phronesis*).

Gadamer implicitly shows on which side of this question he stands in a 1982 letter of reply, which Bernstein has published as an appendix to his book. It contains one of Gadamer's most sustained expressions of confidence in the unconditional possibility of solidarity:

> Clearly your decisive argument is the collapse of all principles in the modern world, and I certainly agree with you that, if this were correct, my insistence on *phronesis* would be nothing more than declamation. But is this really the case? . . . Here, in fact, my divergence from Heidegger is fundamental . . . If it were the case that there were no single locus of solidarity remaining among human beings, whatever society or culture or class or race they might belong to, then common interests could be constituted only by social engineers or tyrants, that is, through anonymous or direct force. But have we reached this point? Will we ever? I believe that we would then be at the brink of unavoidable mutual destruction . . . I am concerned with the fact that the displacement of human reality never goes so far that no forms of solidarity exist any longer. Plato saw this very well: there is no city so corrupted that it does not realize something of the true city; that is what, in my opinion, is the basis for the possibility of practical philosophy.
>
> The conflict of traditions we have today does not seem to me to be anything exceptional.[71]

---

68 Bernstein, *Beyond Objectivism and Relativism,* 261.
69 Bernstein, *Beyond Objectivism and Relativism,* 157.
70 Bernstein, *Beyond Objectivism and Relativism,* 157-167 passim.
71 "Appendix. A Letter," 263-264. As seen in Chapter Three, Gadamer makes a similar point in his 1984 interview: instead of inferring, as Heidegger does, that we

Here Gadamer clearly states his view that no historical situation, including our own, can utterly destroy all forms of solidarity. This conclusion seems to reflect simply his interpretation of the human world; no deeper or more ultimate basis for it (or more practical implication from it) is asserted. His confidence in the possibility of solidarity appears not to be weakened by whatever doubts or incredulity the modern age has about the universals through which solidarity has been traditionally articulated and applied.

Initially, however, it is unclear whether this confidence means that Gadamer thinks traditional universals have not lost as much currency as Bernstein thinks, or whether he thinks the possibility of solidarity does not finally depend on such traditions but on something else. In fact this ambiguity recurs throughout Gadamer's work. On the one hand, the very idea of philosophical hermeneutics only becomes possible, as Gadamer acknowledges in several places, because our age no longer presumes the validity of received traditional universals. For example, he says that "only when our entire culture for the first time saw itself threatened by radical doubt and critique [engendered by the Enlightenment] did hermeneutics become a matter of universal significance."[72] On the other hand, however, as indicated in the beginning and end of the extended quotation above, Gadamer does not share the view that human traditions can be so corrupted that there is nothing left on which we can rely in order to guide us toward solidarity. Indeed, throughout his work, Gadamer's constant reminder to his readers is that traditions have greater power and endurance than we realize. Bernstein also perceives this ambiguity and formulates the problem it implies:

> At the heart of Gadamer's thinking about *praxis* is a paradox. On the one
> hand, he acutely analyzes the deformation of *praxis* in the contemporary

---

must try to endure through a 'world night,' he assumes that there always remain bases of agreement larger and deeper than our areas of disagreement and misunderstanding. "I agree with Plato, who said that there is no city in the world in which the ideal city is not present in some ultimate sense." And more emphatically: "I think that without some agreement, no disagreement is possible. In my opinion, the primacy of disagreement is a prejudice. Beneath the structures of the opinion-making technology on which our society is based one finds a more basic experience involving some agreement." ("Gadamer on Strauss," 9, 10.)

[72]   *RAS*, 100. See also *TM*, xxxiv [xxii].

world, and yet on the other hand he seems to suggest, regardless of the type of community in which we live, that *phronesis* is always a real possibility.[73]

Gadamer's apparent answer to this ambiguity is evident in a passage at the end of "What is Practice?" There Gadamer rhetorically asks whether the "things held in common," including traditional principles about solidarity and practical good in general, though often "beyond the explicit consciousness of anyone," are sufficient grounds for hope in the realization of social and political good. He points our attention to several contemporary issues which are often mentioned as reasons for anxiety or even despair about the destiny of human life in a technological world: genetic breeding possibilities, the flattening of democracy to photogenic contests, the medical postponing of death, and ecological crises.[74]

Gadamer's first response to the threat to solidarity posed by these problems is the very fact that society perceives them as crises. He reminds us of the "shock wave" felt around the world at the thought of genetic breeding; the horror expressed over Stalin's show trials of brainwashed victims; the concern generated over the fate of democracy in post-war consumer society; and the courage of doctors who accept the death of patients. "In the light of these experiences everyone can become expressly aware of the limits of manipulative capacities."[75] Thus the prospect is not completely bleak, in his view, because time and again, in social or individual manifestations, visceral reaction to and concern about the problems we face indicate that a fundamental human capacity to judge wisely has not been destroyed. Each of the above issues, he appears to mean, evokes a response that is not merely an emotional conditioning but the inchoate response of *phronesis*. "Maybe this [societal pes-

---

[73] Bernstein, *Beyond Objectivism and Relativism*, 158. Bernstein himself avoids such a paradox by arguing that we should focus on the "material, social and political conditions [that] need to be concretely realized in order to encourage the flourishing of *phronesis* in all citizens." He acknowledges that *phronesis* can always occur, even in the worst societies, but challenges Gadamer to describe these conditions. (Bernstein, *Beyond Objectivism and Relativism*, 157, 158.) Obviously, questions of causality in regard to *phronesis* or solidarity are important, and fall under the participatory function of practical philosophy. But when the search for the grounds of good practice becomes focussed on causal considerations, as it tends to be for Bernstein, we should by now recognize that practical philosophy is being changed into a form of technology, or as Gadamer puts it in his letter, the classical transition from ethics to politics "now becomes the transition from practical philosophy to social science." ("Appendix. A Letter," 262.)

[74] *RAS*, 82, 83.

[75] *RAS*, 82-84 passim.

simism about the survival of common human values] is too negative a perspective. Perhaps the normative character of practice and hence the efficacy of practical reason is 'in practice' still a lot greater than theory thinks it is."[76]

In the closing passages of his essay, Gadamer seems to admit that this resilience of practical reason is no guarantee of solidarity or of good practice. Even so, he remains carefully yet undeniably hopeful, and it is the reason for his hope which particularly interests us in what follows:

> We are still a far cry from a common awareness that this is a matter of the destiny of everyone on this earth and that the chances of anyone's survival are as small as if a senseless attack with atomic weapons of destruction were to occur if humanity in the course of one or perhaps many, many crises, and in virtue of a history of experience involving many, many sufferings, does not learn to rediscover out of need a new solidarity. No one knows how much time we still have. But perhaps the principle is sound; for reason it is never too late . . .
>
> Perhaps one finds this a rather sad consolation. But I do not mean that this is all: it is but a beginning, an initial awareness of solidarity. Merely out of necessity, to be sure. But is that a real objection? Does it not rather say something for the availability of a *fundamentum in re*? Even a solidarity out of necessity can uncover other solidarities. Just as we . . . are blind to stable, unchanging elements of our social life together, so it could be with the reawakening consciousness of solidarity of a humanity that slowly begins to know itself as humanity, for this means knowing that it belongs together for better or for worse and that it has to solve the problem of its life on this planet. And for this reason I believe in the rediscovery of solidarities that could enter into the future society of humanity.[77]

Having already shown that he takes seriously the evil, ignorance and pretense that *phronesis* faces in a technocratically oriented world, Gadamer clearly wants to show in this passage that he does not despair. Yet his affirmation is not one of blind hope or utopian idealism. In fact it is very modest. Many of his words are about the possibility of solidarity "out of necessity," in view of present or potential threats to our sense of what being human means. It is evidence of a realism and perhaps a shrewdness to say that we may at first 'back into' solidarity out of less than altruistic motives.

It must be pointed out, however, that human responses to stress, suffering or danger often do not bear this out; the historical record fre-

---

[76] *RAS*, 83.
[77] *RAS*, 85-86.

quently does not support Gadamer's hope. Yet, as Gadamer says, "I do not mean that this is all." Finally the basis for his confidence in the possibility of solidarity appears to go beyond what we can dredge from a realism about human nature or human history. The strong implication here is that the basis does not ultimately lie in human hands or efforts but in a "fundamental reality." Gadamer thinks that the "things held in common" are adequate—to sustain our hope for solidarity; to bring solidarity into being in human practice; to serve as resources that cannot be exhausted, even when we do not keep a tight 'hold' on them. These commonalities include the results of our human efforts, but apparently go beyond to include realities we do not control and in which we already find ourselves participating.[78]

From this perspective, the ambiguity of Gadamer's position about the prospects for solidarity is more apparent than actual: he is implicitly asserting that traditional universals, weakened as they may be, still survive and will always survive, but at the same time is also implying that the possibility of solidarity does not causally depend on their survival. Rather, the possibility of solidarity, while it clearly involves human agency for Gadamer, finally depends on a "fundamental reality," and it is because of this reality that traditions of solidarity will never be utterly lost. One consequence of this condition, though Gadamer does not suggest this, is that we can relinquish our notions of being able to control and direct our being toward some distant good, and instead let ourselves participate in the possibility of good which is already present. Therefore—and this is a critical matter on which every ethic or practical philosophy must decide—questions concerning the causality of and social conditions for solidarity are important ones, but only penultimately: they can only be authentically addressed on the basis of this confidence, and not as the means to it. On what may be the most basic decision in practical life, Gadamer's implicit choice reminds us of a path which the technocratic mind can only barely recognize.

Unfortunately, Gadamer's explanation of this confidence is simply too latent and too vague to orient us any further toward what is good. What we have just concluded is the limit to which we can pursue a matter about which Gadamer is hardly forthcoming. It is as though at the brink of systematically relating his confidence to human agency, to moral principles and to social conditions for the realization of solidarity,

---

[78] Gadamer elsewhere appears to indirectly support this interpretation when he writes that the "insight" into the adequacy of these commonalities "has been covered over in our day by historicism and all the varieties of relativist theory." (*RAS*, 82.)

Gadamer chooses to remain silent, as he has before other tasks relevant to the participatory function of practical philosophy. Our interpretation of Gadamer's confidence in the possibility of solidarity attempts to make more explicit the participatory claims inherent in Gadamer's work. But if this confidence is to be something in which others can share, it must be articulated further.

The Conditional and the Unconditional

In the final text of Gadamer's to be examined in this chapter, we return to the type of material which primarily concerns the reflective function of practical philosophy. We come last to this work in some exasperation, having already examined the most promising texts in which Gadamer might speak more substantively and concretely. We also turn to it last knowing that Gadamer here further addresses—albeit from the reflective perspective—the nature of practical philosophy, and hoping that we may nevertheless find material which contains participatory implications—or which might explain why he seems to omit consideration of them.

The text at hand is an essay, "Über die Möglichkeit einer philosophischen Ethik" ("On the Possibility of a Philosophical Ethic"), which was first published in 1963. The general subject in this text is not new to our investigation, although this treatment of it is different in several respects: his focus is the relationship between theoretical assertions of ethical norms and practical ethical action. His attempt to deal with this subject here was his first after *Truth and Method* appeared, and is his most sustained attempt to do so, even when compared to later essays in *Reason in the Age of Science.*[79]

To be more specific, the basic question posed in this essay is whether we can reconcile the unconditionality of moral good with the conditionality of human existence in a way that will make clear what is the role and relevance of practical philosophy to practical life. Actually, Gadamer does not speak of practical philosophy in this essay, but instead uses the

---

[79] This essay has not yet been published in English translation, and barely any discussion of it was evident until recently. Finally, however, it has begun to receive the critical attention it deserves. This essay is referred to by the following two works, especially the latter, both of which are discussed in our final chapter: David Ingram, "Hermeneutics and Truth," in Hollinger, Robert, ed., *Hermeneutics and Praxis* (Notre Dame, IN: University of Notre Dame, 1985), and Ronald Beiner, "Do We Need a Philosophical Ethic? Theory, Prudence, and the Primary of *Ethos*," *The Philosophical Forum* 20, no. 3 (Spring 1989): 230-243.

term 'philosophical ethics,' one he rarely uses elsewhere. He is not very precise about the scope or content of this term in the present essay, but it appears to mean roughly the same as practical philosophy; perhaps at some point after 1963 he decided practical philosophy was the preferable term. Thus, although there may be reasons not to equate these terms, for present purposes we follow Gadamer's apparent equation of them.[80] Our interest in examining this essay, however, is slightly different from Gadamer's: there is enough evidence that the kind of reconciliation he seeks is possible; the question is how Gadamer will articulate it, and how far he will go—or thinks we can safely or usefully go—to articulate it.

As in certain texts discussed in Chapter Two, Gadamer formulates the issue facing philosophical reflection on ethics in terms of the irrelevance (perceived and experienced) of theoretical knowing for practical doing:

> Kierkegaard showed that all 'knowledge at a distance' is not sufficient for the fundamental moral and religious situation of men. As is the case with the meaning of Christian revelation, as 'simultaneously' experienced and heard, so also ethical choice is not a matter of theoretical knowledge, but rather of the brilliance, sharpness and distress of conscience. All knowledge at a distance threatens, by concealment or by weakening, the existing challenge in the ethical situation of choice . . .
>
> In view of such circumstances, philosophical ethics indeed appears to be in an irresolvable dilemma. The generality of reflection—in which, as philosophy, philosophical ethics is necessarily located—becomes entangled in the questionability of all law-like ethics. How should this generality of reflection do justice to the kind of concreteness with which conscience addresses a situation with the feeling of justice, with the reconciliation of love?[81]

---

[80] For our part, we suggest that philosophical ethics should be viewed as a part of practical philosophy. Like Gadamer, we view all kinds of practice as actions which involve moral choices. But practical philosophy should also be concerned with all the dimensions and fields of *praxis*, and its relationship to *techne* and *episteme*, to use Aristotle's terms for the other two major branches of knowledge. By contrast, the scope of philosophical ethics is conventionally limited to discerning how moral choices are or ought to be rationally ordered, either immediately, in terms of normative ethics, or more fundamentally, in terms of meta-ethics.

[81] Hans-Georg Gadamer, "Über die Möglichkeit einer philosophischen Ethik" (originally 1963), in his *Kleine Schriften*, 4 vols. (Tübingen: J. C. B. Mohr, 1976), 2nd ed., 2: 181. Translation of this and other passages quoted from this essay was made by Susan Schultz and this writer.

Gadamer asserts that fundamentally there are only two philosophical paths by which this dilemma can be addressed: either through Kant's ethical formalism or through Aristotle's ethic based on the mediation of ethos and logos. Although he states that neither ethic alone is sufficient but needs the other to be adequate, it is clear that he regards Aristotle's ethic as the more resourceful of the two paths, precisely because it takes account of the conditionality or historicity of the theoretical and normative claims of human reasoning.[82]

> Thus the core of Aristotle's philosophical ethic lies in the mediation between logos and ethos, between the subjectivity of knowledge and the substantiality of being . . . It has been correctly stated that Aristotle's last word on what is right consists of the indefinite formula of 'what is fitting and appropriate' . . . The actual content of the Aristotelian ethic is not found in the great leading conceptions of a heroic ethics of example and its 'table of values,' but in the Insignificance and Infallibility of the concrete moral consciousness . . . which get expressed in such unspeakable and all-encompassing conceptions as what is 'proper,' what is 'decent,' what is 'good and correct.'[83]

In effect, Gadamer's choice of Aristotle's path reflects his affirmation that the polarity of theory and practice is more an appearance than the reality, since the "dilemma" of ethics can be addressed by the reciprocity of theory and practice. Gadamer does not specifically invoke that principle in this essay, but it is certainly implied by the interpretation of Aristotle illustrated above. Moreover, Gadamer describes Aristotle's concept of ethos as one which recognizes "that 'virtue' does not exist only in knowledge," but that such knowledge "depends on 'how' one is," which is "formed by one's upbringing and way of life."[84] This is what defines the "concretization of ourselves" as agents who seek to do and be good, a process which is so particular that Aristotle's ethics intentionally only present an "outline" of that process and its goals.[85]

Gadamer provides barely this, however. He briefly discusses one specific virtue, and in view of our discussion of solidarity in Chapter Four, it is not surprising that this virtue is friendship. The virtue of friendship serves for Gadamer, as it does for Aristotle, as the central link between ethical and political inquiry. Aristotle's concept of good practice

---

[82] "Über die Möglichkeit," 181, 186-189.
[83] "Über die Möglichkeit," 187.
[84] "Über die Möglichkeit," 186.
[85] "Über die Möglichkeit," 188.

is referred to by Gadamer as what is feasible, practicable, or simply 'do-able.' It is developed in terms of a coordination of deontological and teleological perspectives. The following is his conclusion about the relationship between these classic alternatives in ethical theory:

> The practicable—of course this is not only that which is right, but also what is useful, goal-oriented, and to this extent is 'right.' The permeation of both these 'right-nesses,' as manifested in the practical behavior of men, constitutes the human good for Aristotle. Certainly man does not engage in morally correct behavior in the sense of being goal-oriented like the artisan who knows his business (*techne*); moral behavior is not made right just because what has been produced is right, but rather its rightness lies also and above all in ourselves, in the How of our behavior—in exactly how it is done by the man who 'has it right' (the *spoudaios aner* [good man]).[86]

In the penultimate passage of this essay, Gadamer summarizes his perspective on ethics, a perspective which represents the fruit of a 'fusion' of his 'horizon' with Aristotle's:

> Precisely . . . the conditionality of our insight in general—wherever it does not involve decision in the eminent sense of the word but rather the choice (*prohairesis*) of the better—is not a defect and not a barrier: it has its positive content in the social-political concrete definition [*Bestimmtheit*] of the individual. This definiteness is however more than a dependency on the changing conditions of social and historical life. Certainly everyone is dependent on what is presented by his time and his world, but neither the legitimacy of moral skepticism nor the technical manipulation of all opinion-formation by the use of political power follow from this. The changes which grip the customs and way of thinking of an age, and which tend to give particularly the elderly the threatening impression of a total dissolution of morality, take place on a still foundation [*ruhende Grunde*]. Family, society and state define the essential composition of men insofar as they fill his ethos with changing content. Of course no one can say all of what will be possible for men and their way of living together—and yet this does not mean that everything is possible, that everything can be organized and established arbitrarily and by whim, as the powerful would like. There is a right according to nature. To the conditionality of all moral knowledge which pervades moral and political being, Aristotle finds a counter-balance in the conviction, shared with Plato, that the order of being is strong enough to put limits on all human confusion.[87]

---

86 "Über die Möglichkeit," 188. Transliteration and translation from Greek added.
87 "Über die Möglichkeit," 191.

This passage is noteworthy in more than one respect. It effectively corroborates the general interpretation of Gadamer's perspective on practice as presented in Chapters Three and Four. It is not in abstract theory but in the concrete human ethos, in the historical scene of human living, that we find the communicable "content" of ethics: every agent is defined, in a sense even determined, Gadamer says, by his or her socio-political situation, by family, society and state. The point being made is not any metaphysical conclusion about arbitrary relativity or about any form of determinism. Rather, Gadamer wants to reinforce in our minds the limitations of theoretical judgment about practical life in general, and encourage us to develop and rely on *phronesis*. Ronald Beiner correctly emphasizes this interpretation of Gadamer's essay in what is one of the few published works to deal with this essay.[88] As he points out, "theory as such is incapable of anticipating" the infinity of circumstances "which face us in concrete life"; "this means that it [theory] stops short of precisely that concreteness that is at the heart of all moral knowledge."[89] For this we must rely on *phronesis*. Beiner goes on to add this helpful commentary:

> In order for moral convictions to have force within the life of concrete societies, ethical intuitions must possess a great deal more self-certainty than they could possibly gather from merely theoretical demonstrations . . . One must act *as if* unreflectively, embodying a sure sense of what is good and right; one must command a kind of practical assurance that even the strictest, most rigorous set of arguments fails to supply. This is something made available only by character and habituation, never by rational argument as such. As an Aristotelian would say, in order to live virtuously and to make the right choices, one's soul must be shaped by certain habits of virtuous conduct, in a way that renders superfluous recourse to strict arguments.[90]

But let us return to the text on which Beiner is commenting, for Gadamer has more to say, which Beiner strangely ignores. He goes beyond simply denying that the utter conditionality (or historicity) should be equated with moral skepticism or arbitrary relativism. Instead, he

---

[88]　Ronald Beiner, "Do We Need a Philosophical Ethics? Theory, Prudence, and the Primacy of *Ethos*," *Philosophical Forum* 20, no. 3 (Spring 1989): 230-243.

[89]　Beiner, "Do We Need," 234.

[90]　Beiner, "Do We Need," 235-236. Beiner also offers this aphorism to summarize his characterization of Gadamer's view: "Good theory is no substitute for good socialization, and even the best theory is utterly helpless in the face of bad socialization." (237)

makes what amounts to an ontological assertion: the very conditionality of human life rests on a "still foundation," on the "order of being," which is more powerful than the disorder and perplexity that accompanies our conditionality. Taken with the above point, we have another and more direct confirmation that Gadamer's perspective is an affirmation of the compatibility of historicity and good.

This is particularly evident in the essay's final paragraph, which, however, Gadamer poses in an uncharacteristically enigmatic fashion. He states the fundamental principle or criterion of philosophical ethics, which he has learned from Aristotle and which answers the question of how philosophical ethics is possible in view of its historicity:

> Aristotle acknowledges the conditionality of all human being in the content of his teaching about ethos without this teaching denying its own conditionality. A philosophical ethic which not only knows about its own questionability but precisely has this questionability as its fundamental content—only this appears to me to be sufficient for the unconditionality of what is moral.[91]

This provocative formulation of the criterion of a philosophical ethic, or a practical philosophy—a criterion which can be called his 'questionability principle'—requires some attention. "Questionability" (*Fragwürdigkeit*) here seems to have a meaning closer to 'contestable,' than to 'refutable.' Thus Gadamer's principle appears to mean that a philosophical ethic should know about its own contestability and should acknowledge that very contestability as an essential part of its content. Gadamer is clearly trying to say something about the dialectical character of ethical knowledge and not merely make an analytical statement. In fact, it does not make analytical sense to say 'questionability is the content of philosophical ethics,' or that 'the content is questionability and this content is questionable.' But it does make dialectical sense to say that a substantive content asserted by philosophical ethics should be subject to questioning—and that awareness of the contestability of all such content is itself a corollary to the realization of what is unconditional.

How, then, do Gadamer's statements in this essay bear on the participatory function of practical philosophy? As several of the passages indicate, Gadamer recognizes the moral universals found in our traditions and histories as having an essential role in moral reflection and delibera-

---

[91] "Über die Möglichkeit," 191.

tion.[92] Since these are the kinds of resources with which the participatory function of practical philosophy is concerned, it is reasonable to infer that Gadamer implicitly supports the idea that practical philosophy has such a function. To put the matter differently: if Gadamer thinks the contents of our ethos need not or cannot be made available to us through theoretical reflection, he is mistaken. Yet he implies precisely this, as though ethos had nothing to do with theory, although the reciprocity principle of his own hermeneutical perspective suggests otherwise.

As the questionability principle indicates, if an ethic (or a practical philosophy) is to show what is morally unconditional, it must be ready to have its content contested; but it must, therefore, *have* a content, and it is this which Gadamer does not provide or even outline. That is, Gadamer has not provided any substantive ethic which, according to his own questionability principle, must be articulated and allowed to be contested. He has apparently chosen not to address the participatory function of practical philosophy—the function which would articulate those contestable claims—even as he also presupposes its necessity. As a result, the significance of the questionability principle lacks concrete illustration and remains elusive, as Gadamer only presents the abstract result of his dialectical perspective without leading us through the interdependent relationship between the two functions of practical philosophy.

A similar point can be made with regard to Gadamer's comment on deontological and teleological views in normative ethics. From the perspective of the reflective function, he makes the salient point that the good is characterized by a "permeation of both these 'right-nesses,'" which is concretized "in the How of our behavior."[93] It may be of little import to Gadamer that his own normative ethic remains unarticulated, but it cannot be immaterial to him that the practical relevance of his reflective point remains clouded without this articulation. Gadamer's focus on the How of practice corresponds to the reflective function, and the purpose of teleological and deontological theories corresponds to the participatory function: the interdependence of these functions means that Gadamer's point will itself become more vivid when it is seen through the substance of a normative ethic.

---

92 In his letter to Bernstein, Gadamer restates this point: "Practical philosophy insists on the guiding function of *phronesis*, which does not propose any new ethics, but rather clarifies and concretizes given normative contents." ("Appendix. A Letter," 262-263.)

93 "Über die Möglichkeit," 188.

This relationship of the reflective and participatory functions is overlooked earlier in the essay as well. Gadamer refers to Aristotle's ethic as providing an "outline" of what the "concretization of ourselves" means and contains.[94] It is clear that what he refers to there as an outline is equivalent to Aristotle's participatory content, a content about virtues, society and institutions, in which the principle of Aristotle's philosophical ethic—the mediation of logos and ethos—would be recognizable. Except for a brief reference to the virtue of friendship, Gadamer prefers to emphasize precisely that it is an outline, no more than a rough sketch, and de-emphasizes its content and its role as something which, though still theoretical, is located between abstract philosophical principles and the actual "concretization of ourselves." But if, as Gadamer seems to think, Aristotle's outline contains valuable insights into the general nature of practical good, then it would be important to provide the kind of hermeneutic of that outline (i.e., Aristotle's participatory content) which would make it possible to more concretely demonstrate the mediation of logos and ethos in today's world. Moreover, such a 'fusion of horizons' would complement and complete the hermeneutic Gadamer has already made of the reflective content of Aristotle's philosophy of practice.

By stopping where he does, Gadamer exposes his practical philosophy to a problem he perhaps had not anticipated. In the penultimate paragraph of the essay, quoted above, he says that the "insight" on which we base our practical decisions "has its positive content in the social-political concrete definition of the individual."[95] But little else is said about which contents he means, or about which of those contents are, in his view, most deserving of moral attention or practical application.[96] To be silent about making discriminating judgments of this kind amounts to approving of all practices simply because they are based on—or even causally determined by—the time and place in which one lives. We ascertained in Chapter Four that philosophical hermeneutics is not contextually relativistic, and Gadamer himself warns that "not . . . everything is possible."[97] However, by providing virtually no substantive target of the

---

[94] "Über die Möglichkeit," 188.

[95] "Über die Möglichkeit," 191.

[96] It can also be noted, however, that later, in other essays, he at least introduces the concept of solidarity. Although he speaks of few other norms, and even this one is not much explored, it is an advance beyond the limit at which he stopped in 1963.

[97] "Über die Möglichkeit," 191.

good himself, he leaves his practical philosophy open to the accusation of contextual relativism.

REACHING GADAMER'S LIMIT

We began this survey of Gadamer's practical philosophy with a conception of the two functions—reflective and participatory—by which practical philosophy fulfills its purpose of targeting what is good. Much of the work by Gadamer which we then examined in this chapter is, as in previous chapters, devoted to practical philosophy's reflective function. Gadamer's views in this regard are persuasive, and include further demonstrations of his confidence in the compatibility of historicity and good.

But his attention to the participatory function of practical philosophy remains less than adequate—and certainly less than was possible, given the opportunities he accepted.[98] To be sure, there are hints and traces of participatory claims. We have seen his claims for the value of solidarity, his critique of modern practice, and his attention (such as it is) to the question of whether solidarity is the precondition as well as the hoped-for effect of good practice. Several of these discussions take us into new territories, of which there is no hint in the more abstract passages of his works which were largely our focus in preceding chapters. More strikingly, some of these discussions—notably his 1965 address, "Notes on Planning the Future," and his 1963 essay, "Über die Möglichkeit einer philosophischen Ethik"—appeared *before* his debate with Habermas, and therefore without the widely-noted provocation which that debate created. Clearly, while that debate precipitated some, it did not precipitate all of his public reflections on modern practice and the role of reason in good practices. In fact, the material surveyed here further establishes that there is a body of Gadamer's work on these subjects which stretches from his studies of Plato's ethics and politics in the 1930s[99] to his most recent essays.

---

[98] Gadamer does provide a few other discussions—usually even more brief and cursory—of potentially relevant topics. But from what is known to this author, the texts we have examined are representative of the scope and manner in which Gadamer addresses the questions we have posed, and a longer survey seems unlikely to secure results of any significant difference.

[99] See: Robert Sullivan, *Political Hermeneutics*, esp. Chapter Five.

Yet, with regard to the participatory function of practical philosophy, the material seen in this chapter does not embrace the responsibility which, from the hermeneutical point of view, such a philosophy implies. The notable points to which we have brought attention in our survey are encouraging and sometimes intriguing, but too often unsatisfying: they are vague, or over-generalized, or merely hints—features which, in a practical philosophy, are different from a content which is simply and properly general.[100] We encounter grand visions left unexplored and enigmatic principles unexplained. Conflicting statements are not unknown. These texts have obvious merit; but they barely begin to fill the evident lacunae of his practical philosophy when it is viewed as a whole. He articulates no theory of society or power, no investigation of justice or friendship, no clear concept of moral anthropology, no vision of human purpose or identifiable philosophy of history, no religious or theological perspective, and no political philosophy or program. The participatory function does not, of course, require that all of these projects be undertaken in their entirety, nor that all his ethereal generalities be condensed into more tactile substance. But given the purpose of practical philosophy, which Gadamer himself has elucidated and defended in general outline, his silence on such subjects draws objection as well as attention.

Upon overall reflection, it seems as though Gadamer emasculates his practical philosophy at the moment he most elevates it. Perhaps, when shorn of his ambiguous and fragmented claims, Gadamer's theory boils down to the claim that 'close attention to the hermeneutical character of all claims (including his ontological and moral claims) will, in itself, help us identify the participatory content which best targets what is good here and now.' We accept this principle; hardly recognized, it is perhaps near the pinnacle of his hermeneutical philosophy of practice. But we go on to point out that Gadamer has not used it. He simply stops there, seeming to think that enunciation of this principle—his hermeneutical ethic, we might call it—is enough to satisfy the core requirements of a practical philosophy. It is not simply that living and breathing people need more than this to justify their existence. And it is not just that people are more vocal and expressive about their environment than this. It is also that

---

[100] Much earlier, we saw Gadamer himself uphold an even higher standard when, in agreement with Aristotle, he says that "a science of the 'good in general' . . . is meaningless for practical philosophy." (*The Idea of the Good*, 160.)

there is simply more to be said. Yet Gadamer knows all this—which makes it more perplexing that he stops.

Or perhaps Gadamer's hermeneutical ethic is even more specific; maybe he means to say something like this: 'In the first place, philosophical ethics or practical philosophy only teaches the form of moral knowledge, not its substance—a form revealed to the hermeneutically sensitive consciousness. Yet, in the second place, even this form is itself genuine substance, is something applicable and good to apply.' If this is the case, Gadamer's practical philosophy does not just refer us elsewhere to find the substance of an ethic but, having surveyed many elements of tradition, asserts at least this one substantive claim; that is, it is devoted to elaborating what the form of moral knowledge is that is itself a substance worth applying.

If this is Gadamer's meaning, we have already agreed with the second statement posed above: at this level of reflection, form is indeed part of content. But partly for this reason, we cannot agree that practical philosophy should only concern itself with this point, as the first statement suggests. There is no cause for a hermeneutical ethic to avoid the content or substance of moral claims in supposed deference to the greater authority of their (hermeneutical) form. In fact, we have seen reasons why such an ethic cannot be said to be really aware of its hermeneutical character until it also acknowledges that it is not, and cannot be, aware of all it is asserting. We are more likely to acknowledge this when experience reminds us that the formal assertions of a hermeneutical ethic are bound up with living contents, which both inundate us from without and burst from inside us until we can no longer track all our intentions.

As it is, Gadamer essentially gives us a practical philosophy in which we are instructed about only one of its two functions. This is more than half a practical philosophy only in the imagination of one who still thinks the summits of theory are the highest. It is less than half a practical philosophy to the practical agent each of us hopes to be. The 'conception' of a hermeneutical philosophy of practice must refer to both the formal concept of its nature and functions, and to the creation of its substance and content; in Gadamer's case, the near absence of the latter suggests a mis-application of the former.

We do not mean that by filling this gap, the theory-practice reciprocity would be fully incarnated. Obviously, we are only speaking from the theoretical side about this reciprocity; the practical demonstration of this reciprocity must itself be practical. But Gadamer has not remembered this reciprocity while he theorized, and therefore has not incorporated into his work the interdependence of the two functions of practical philosophy. In fact, he has failed to adhere in his own work to the very concept—that of the application of prejudice—which notably distinguishes his own position from Strauss' and Habermas', and, as we have shown, makes his position in those debates the better one.

Gadamer's own language helps make the point: he holds before us the sobering reminder that, in practice, there are 'no rules for applying rules,' no meta-rules for applying ordinary rules. But a practical philosophy which only tells us this has not really gotten started: it does not articulate the 'ordinary rules' whose application may indicate the direction in which the target will be found. We must conclude that wherever Gadamer becomes silent about such things, we have reached the limit of his direct relevance to the development of a hermeneutical approach to practical philosophy.

SIX

Gadamer's Silence

Gadamer's curious termination of the development of his practical philosophy raises questions about how one might go further, and why he chose a path which stops prematurely. There is no direct acknowledgement of, much less explanation for, the cessation of his labors in practical philosophy when they reach some line still visible only to him. He does not indicate in which directions he thinks this project could be pursued by others; in fact, he seems to doubt that it could be further pursued in any significant way.

Positioned behind these questions, however, is the hesitant but basic assumption of our dialogue with Gadamer that all along he was in fact forming a practical philosophy, or at least the core of one, and not merely standing on the side to tell us about that branch of philosophical reflection. This is certainly what common sense tells us after reading all that Gadamer has to say on the subject; yet, whether Gadamer actually thinks so is, once more, not as clear.[1]

Nevertheless, Gadamer seems to have some vision or outline of a practical philosophy in mind, a body of thought for which he has formulated what he thinks are the central principles, but which is perhaps still in the process of formation, of fusing new horizons with old traditions.

---

1 If it were insisted that Gadamer was instead only telling us about practical philosophy, then our dialogue with Gadamer in Chapter Five and in this chapter would have to take on a moderately different character. However, the constructive goal of the inquiry—the clarification of a hermeneutical philosophy of practice—and the net result of the dialogue with Gadamer as it bears on that goal, would remain the same.

And while he would quickly agree that such a labor is not easy in our age, which keenly feels its discontinuity with traditions once taken for granted, he evidently concludes that the task is not impossible nor even overly daunting, and is rather simply the difficult work we must do.[2] At any rate, it does seem a reasonable extrapolation from our sources to assume that Gadamer is open to and indeed would welcome the further development by others of practical philosophy as he envisions it.

Of course, this last qualification—'as he envisions it'—raises a critical point, and one on which Gadamer is eminently vague. While one cannot imagine him refusing to share or release his work into the hands of others, he clearly does not want to see his work used either simply as a storehouse of neutral methods and procedures, or as a foundation for some specific ethic, political philosophy or social policy program. Up to some point, this is both philosophically reasonable and practically prudent as a means to protect his ideas from improper use. Now, most philosophers use their work to point in the direction of either a neutral method or committed ethic—or even (and perhaps preferably) toward both. But Gadamer appears to wish his work did not point in either direction. This would be a 'purity' that is indistinguishable from irrelevance. More importantly, abstention from all pointing is not an option in human practice. In short, Gadamer cannot hold his nose and close his mouth as well. The consequence for us is that the way forward is open, but not clear; with regard to the queries posed above, we are left (quite appropriately) to our own devices.

In order to arrive at answers, the goal of deciphering Gadamer's intentions does not matter as much as discerning the direction in which his philosophy points, even if this is different from the direction in which its author moves. And consequently, the first criterion by which to pursue our final questions is to apply philosophical hermeneutics itself: to inquire whether his silences or someone else's assertions better conforms to the direction in which the hermeneutical ontology points. We may have reached the limits of Gadamer's relevance for practical philosophy, but it is worth wagering that we have not reached the end of hermeneutical philosophy's relevance.

The two questions on which this chapter will focus, then, are these: First, in what directions can practical philosophy be developed beyond the point at which Gadamer stops? Since we have already shown how

---

2 For example, Gadamer's reply to Bernstein, quoted in Chapter Five: "The conflict of traditions we have today does not seem to me to be anything exceptional." ("Appendix. A Letter," 264.)

Gadamer has not adequately addressed the participatory function, our focus will be on the forms in which this lacunae might be filled. Second, why does Gadamer stop where he does? We may be allowed some curiosity about the man, but the significance of this question bears with greater weight on our larger substantive concerns—which include discerning how future efforts might avoid any mistakes of Gadamer's. It is appropriate in this work to give our greater and final attention to the second topic, to Gadamer and his relationship to what he gives to (and leaves for) his readers.[3]

Since hermeneutical philosophy itself serves as a criterion of adequacy, our critical and constructive efforts here should be distinguishable from the objections and projects of Leo Strauss and Jürgen Habermas. It goes without saying that our proposals should also be more compelling to Gadamer, since they do not proceed from premises foreign to his work.

PATHS BEYOND GADAMER

The further development of the participatory function of practical philosophy can happen in either—and very probably, in both—of two directions. They share the general aim of introducing an axiology, and differ with regard to the source being investigated.[4] The effort here to explore these two paths only goes a little distance beyond where Gadamer stops, since the exploration of these paths presumes a scope which exceeds the limits of this work. Our contribution here is to broadly identify the terrain in which such paths can be cut and suggest the fruits that will be encountered there.

Toward the Axiology Implied by Hermeneutical Philosophy

The first path of development focusses on the resources we already present in philosophical hermeneutics. More specifically, we can exam-

---

[3] The order of treating these two topics, while it may seem illogical, has other benefits as well. By first surveying the potential growth of hermeneutically-oriented practical philosophy, we add support to the independent perspective taken in the subsequent inquiry into the reasons for Gadamer's silence. And by creatively addressing the concerns which appear to prompt Gadamer's silence, we offer some new elements to practical philosophy.

[4] By an axiology we will mean not only a philosophical theory of value but also any vision or articulation of values, whether goals or obligations, which purports to frame, focus, or target good practice.

ine the principles of philosophical hermeneutics to see what values can be discerned in them, or can be supported by them. This involves examining Gadamer's work independently of his own intentions for it, of his inferences from it, and of the moral values which may be associated with his philosophy but which in fact he chooses on other grounds. It may then be possible to identify values implied by Gadamer's philosophy which in some way go beyond what Gadamer himself seems ready to allow into practical philosophy.

The first step in this path is to acknowledge that some kind of moral vision is clearly evident in Gadamer's work. Substantively, the norms in question, of course, are those which make up the constellation of terms we have encountered before—solidarity, community, common reality, and judging-with. But philosophically, the first reason for thinking along these lines precedes the evidence of moral values in Gadamer's philosophy. It is the idea that, in principle, theories contain axiological assumptions by virtue of the reciprocity of theory and practice.

It is important to discern, though, whether this moral vision of solidarity is intrinsic to, or can at least be supported by, philosophical hermeneutics—the possibility which concerns us here—or whether it is extrinsic to hermeneutical principles, and represents judgments which Gadamer makes on separate grounds. In the second case, such norms would be allocated to the second path of development, discussed later. In the first case, even evidence that philosophical hermeneutics contains or supports a particular axiology does not necessarily validate that axiology. It does suggest, however, that if Gadamer's basic principles are accepted, the norms which cohere with them should be equally persuasive. Those who closely follow Gadamer must then decide either to redouble their efforts to exclude all axiological elements from hermeneutical philosophy (something which seems impossible), or to modify their views of philosophical hermeneutics and accept that it has moral dimensions of the kind at issue here.

Theoretical reflection in this direction can be advanced through the idea that philosophical hermeneutics contains a teleology, i.e., that it asserts a definite set of values as the purpose of moral action. This notion is not entirely new, and the expression of it most congenial to our inquiry is David Ingram's essay, "Hermeneutics and Truth."[5] While he is not the first to suggest that Gadamer's philosophy is in fact making moral

---

5   David Ingram, "Hermeneutics and Truth," 32-53 in *Hermeneutics and Praxis*, ed. Robert Hollinger (Notre Dame, IN: University of Notre Dame Press, 1985).

claims, Ingram himself observes that the "teleological dimension" of
Gadamer's hermeneutics has been "strangely ignored" by most commentators.[6] Moreover, very few would also assert that Gadamer's moral
vision must necessarily lie at the core of his philosophy.

Ingram describes Gadamer's teleology in this way: "the process of
interpretation 'which we are' is itself teleologically oriented toward a
state of openness (*Offenheit*) and mutual recognition," "which binds [the
participants] together in a common effort to achieve agreement."[7] This
telos of understanding, as Ingram develops it, consists of both openness
and agreement (or "mutual recognition" as he prefers). For as Ingram
reminds us, Gadamer's dialectic is not Hegelian: subject and object do
not become so identified in agreement that openness no longer exists nor
is needed. Rather, Ingram understands Gadamer to mean that no final
telos of human understanding can be identified;[8] instead, all that can be
said along these lines is that

> the maturation of free, self-conscious, universal spirituality is itself impelled
> by a dialogical interplay of prejudices which progressively evolves higher
> levels of openness and reciprocity.[9]

Distilling this into the notion of 'dialogical reciprocity,' Ingram clearly
sees this as the central moral norm and telos asserted by Gadamer's
work, and one to which he himself is also attracted: "might not [this dialogical reciprocity] procure for humanity a *normative* direction which is
not subject to the relativistic vicissitudes of time and place?"[10]

Seeking to find clues about the specific status of this norm, Ingram
notes, as we have, that Gadamer sends confusing signals: on the one
hand, "it is perfectly obvious that Gadamer does attach some prescriptive import to dialogical reciprocity," as seen especially in his debate
with Habermas; but on the other hand, Gadamer also "repudiates the

---

6 Ingram, 40. Another expression of Ingram's general congeniality with our viewpoint is his description of *Truth and Method* as "an attempt to justify a theory of truth
which takes into consideration . . . [a new, hermeneutical concept of] relativity without, however, abandoning the idea of a teleological advance." (41.)

7 Ingram, 39, 40. Ingram goes on to speak of these as constituting a *"meta-
hermeneutical* pre-condition" or a "transcendental condition" of "communicative understanding." (40, 45.) It is not clear what these phrases mean here, nor what they
add; moreover, they are uncomfortably reminiscent of Habermas.

8 Ingram, 43.

9 Ingram, 45. The phrase "progressively evolves higher levels" is confusing since
it implies a progression which may not be appropriate to describe Gadamer's view.

10 Ingram, 45.

possibility of any transcendentally justified ethic."[11] Ingram's interpretation of this tension demonstrates how the notion of a hermeneutical teleology could be developed, and leaves a trail we can also follow, at least in some places.

He considers the model of biological equilibrium, briefly identified by Gadamer in "Notes on Planning for the Future," to have particular significance. Viewed from the hermeneutical perspective, this model draws attention to how social life rests fundamentally on a dialogue among all persons—who are equal participants and also members of a "web . . . of mutual agreement" which "extends beyond the unitary control of particular agencies."[12] From this Ingram goes on to make the kind of societal inferences which Gadamer is reluctant to assert but which are necessary to the functioning of practical philosophy:

> The normative scope of dialogical reciprocity is universal and necessary for all 'true' (i.e., harmonious) [sic] communities, regardless of the specific forms of institutionalized authority and inequality they possess.[13]

Having made this affirmation, Ingram takes steps to account for Gadamer's silence in precisely this area:

> Because the meaning of reciprocity [as the telos of social life] has been extended to cover a seemingly indefinite range of social arrangements, it is doubtful whether any *concrete, prescriptive* content can be generated from it. Gadamer reminds us that this is in perfect keeping with the general tenor of his philosophical hermeneutics, which extrudes methodological and prescriptive considerations in favor of *ontological description*.[14]

However, Ingram goes on to assert that the norm of dialogical reciprocity does indeed have prescriptive implications at an even more specific level than he indicated earlier:

---

11   Ingram, 45, 46. In support of the second quotation, Ingram cites Gadamer, "Über die Möglichkeit einer philosophischen Ethik," 184-188.

12   Ingram, 45.

13   Ingram, 46. Here he illustrates some of the concrete implications of this principle: "The tendency of this line of thought is to treat relations of authority and inequality . . . as epiphenomenal manifestations of a prior consent freely given by those occupying subordinate positions in the social hierarchy. Consequently, . . . [taking the example of social relations in a free market] the rhetorical hegemony which corporate powers apparently exercise . . . is illusory because it is dialogically checked and countered by individual consumers . . . " (Ingram, 45-46.)

14   Ingram, 46. In a footnote, Ingram cites *TM*, [xvi], which appears on xxvii in the 2nd ed.

Dialogical reciprocity is *not* compatible with a state of affairs in which socio-economic disparities and political privileges constrain the 'natural equilibrium' of dialogical checks and balances—*be they acknowledged legitimate or otherwise.*[15]

When Ingram makes this judgment of incompatility, it appears to violate Gadamer's pledge of prescriptive silence, a pledge which Ingram himself just implicitly accepted as valid. But Ingram obviously thinks that Gadamer too has made prescriptive judgments at this level. (This leaves it quite unclear what Ingram makes of this pledge, or of how well he thinks Gadamer has adhered to it.) As evidence of Gadamer's prescriptiveness, Ingram quotes a statement by Gadamer, cited in the Introduction to the present inquiry, which identifies the purpose of *hermeneutical* philosophy—not just of practical philosophy—as being "to defend practical and political reason against the domination of technology based on science," and to vindicate the free responsibility of citizens to decide their common political destiny.[16]

In the face of statements such as these, and in view of Ingram's interpretation of the teleological character of philosophical hermeneutics, Ingram poses two critical questions. In effect, he rephrases, in sharper terms now, the tension which we saw him note earlier between Gadamer's teleology and his pledge of silence. The implications contained in the following questions require careful comment since Ingram fails to make adequate distinctions on issues critical to the development of the path at stake here. Speaking of Gadamer's defense of practical reason, Ingram asks,

Is the above citation not a moral *recommendation* to promote the cause of a more democratic society? If so, how can it be squared with Gadamer's earlier contention that philosophical hermeneutics essentially prescinds from value commitments?[17]

For Ingram, these questions are only rhetorical. With regard to the first, he clearly wishes to join Gadamer in what he takes to be some sort of hermeneutical endorsement of democracy, if he can only figure out what sort that is. The second question is analyzed by Ingram in terms of two possible formulations, with the intention of reconceiving Gadamer's pledge of silence in such a way that it is consistent with his moral

---

15 Ingram, 47.
16 "Hermeneutics and Social Science," 314.
17 Ingram, 47. His italics.

endorsements. First, Gadamer can "construe his recommendations as a particular, culture-bound interpretation," and cast on it the sanction of philosophical hermeneutics; Ingram rejects this option because the authority of a particular tradition, here one which concerns democracy, should not be sanctioned by a philosophy with universal intent. (This option creates confusion; for instance, does Ingram think, or does he not think, that Gadamer is recommending democracy?)

The alternative formulation is one Ingram apparently prefers, and which he also thinks Habermas takes in his own work: to "interpret Gadamer's ontological hermeneutic as axiological in the strong sense of the word, i.e., as implying a *definite* normative content," presumably the content which Ingram identifies as dialogical reciprocity, and which, he thinks, might appear in the form of a "projection of a universal history from a practical standpoint."[18] (Substantively, this option seems distinct from the first only in that Ingram seems to believe it arises from philosophical hermeneutics 'itself,' rather than from any concrete tradition; but this does not seem to be supportable either, in this case because it is contrary to basic hermeneutical principles to conceive a philosophy as independent of all traditions.) Ingram then associates this commitment with Gadamer's brief references to freedom, again implying that this shows a significant convergence or agreement with Habermas.[19] These passages are not very compelling, and neither Gadamer's alleged commitment nor his supposed agreement with Habermas is further clarified. Ingram's boldest statement remains rather hollow: "There is no getting around the fact that Gadamer does subscribe to a teleology which specifies . . . the ideal presupposition and direction of all human understanding. Surely this is tantamount to a transcendental justification of a norm having considerable prescriptive, critical impact."[20]

Consider once again the two questions Ingram posed above. With regard to the first question, Gadamer obviously has made a moral affirmation when he defends the right and responsibility of dialogue in all human concerns. And such an approval depends on an axiology and an interpretation which is inherent in philosophical hermeneutics. Beyond this, it is not difficult to see how such an approval of dialogue supports western democratic values; we, and possibly Gadamer himself, may wish to develop such a connection, or other such connections to the moral traditions we know.

---

18 Ingram, 48.
19 Ingram, 48-49.
20 Ingram, 49.

But tying the knot of such a connection is precisely a step 'beyond' the approval of dialogue. A recommendation of democracy is not a logically necessary implication of the principles of philosophical hermeneutics; it is one moral application of the approval of dialogue, and to be adequately justified it must be tied to other premises as well—premises which are extrinsic to that philosophy. The other end of this connection is also loose: the approval of dialogue may be applicable to democracy as understood in the West, but it may also be applicable to other political orders as well and to other cultures outside of the West, with results we may not be able to imagine yet. Moreover, it is not clear that any one democractic order has achieved very much of what "dialogical reciprocity" implies, even if we prefer that order to others.

Many connections between hermeneutical insight into dialogue and political institutions are possible, and remain ready for development; the implication of hermeneutics is definitely to push and pull us in the direction of making such applications. Ingram contributes to this as he expands the boundaries of Gadamer's practical philosophy, but the teleological dimension of hermeneutical philosophy does not mean that it contains or insists upon specific individual, social or political ethical judgments or ethical systems. Rather, a more careful elaboration will be needed of various kinds of prescriptive norms and of their relationships to basic principles in Gadamer's ontological ontology.

One demonstration of this task concerns a seductive interpretation of Gadamer's work, one which is widely evident, and latently appears in Ingram as well. His essay implies that Gadamer, by acknowledging the approval (or prescriptive implication) of dialogue, is moving—and should move—in the direction of Habermas' critical theory. Recall Ingram's statement quoted earlier, to the effect that Gadamer has come to view dialogue as incompatible with ideology and the inequalities it creates and conceals. Ingram seems to think that recognition of this incompatibility must drive Gadamer in the direction of Habermas. This is an understandable but mistaken interpretation of the prescriptive judgment to which philosophical hermeneutics points. When seen from the perspective developed in Chapter Four, dialogue both is and is not compatible with inequality, depending on the sense and level at hand. It is, once more, the ontological basis of this seeming paradox that sets any practical philosophy based on philosophical hermeneutics apart from Habermas' project, and which starts to dim from view in every effort to 'catch up' with the shiny utility of critical theory. Inequality is inimical to dialogue because inequality imposes handicaps, barriers and deformations on the openness, the mutual recognition (Ingram's terms), on the 'judging-with' (Beiner), on the solidarity (Gadamer) which is being affirmed. But dialogue is not something which can be ultimately deformed—it is a possibility which can be invoked in even the worst conditions of inequality and injustice. In this sense, dialogue 'co-exists' with these evils, and, in fact, it is evil which is compelled to continually rebuff dialogue, instead of dialogue which must fight for breath.

It is gratifying that Gadamer's attention has been brought to bear on the prescriptive practical implication of his philosophy (which Ingram refers to as Gadamer's "recommendation"): it is good to remove inequality and ideology for the sake of dialogue, and in order to remove them we must make judgments about which conditions are better or worse for the fruition of dialogue. But expressing this implication does not alter or qualify the existence of the prior (both in philosophical terms and in the chronology of Gadamer's works) normative practical implication: it is good that dialogue and its analogues are rooted in reality, that they do not depend on our efforts or on conditions we must create or meet, and that they are, in effect, 'there for the taking,' within the reach of every spirit. It would be a practical mistake, as well as unfair to hermeneutical philosophy, to miss this in the name of acknowledging that dialogue is incompatible with inequality or with injustice in general. The advantaged person and the oppressor know the shades of guilt, and the disadvantaged and the victim know the colors of hope, not in spite of the end of dialogue between them, obvious as that end may be, but because their dialogue never does end.

With regard to Ingram's second question, quoted above, it is clear that Gadamer's approval or recommendation of dialogue as a social practice—to say nothing of democracy or other recommendations—cannot be squared with his pledge of silence on such matters, and Ingram agrees. There may be technical utility in separating or circumscribing certain theoretical fields for particular purposes, and Gadamer's pledge of silence might be appropriate in some cases. But when looking at the whole scene, and especially in view of the above discussion, that pledge cannot in principle be justified. What must be added, though, is that the resolution of this inconsistency—between his recommendation of dialogue and his pledge of silence—is to dispose of the latter.

In addition to Ingram's proposal concerning Gadamer's teleology, another possible tracing of this first path can be offered. Ingram suggestively speaks of Gadamer's "reconciliation of ontology and axiology;"[21] he does not explore this notion further but it leads to an even stronger affirmation of the moral significance of philosophical hermeneutics than Ingram presents. We have noted how Gadamer speaks, however briefly and generally, of solidarity and related values as a constellation of moral norms. We have also seen how closely Gadamer associates these norms with communicative events which he describes in terms of dialogue, conversation and rhetoric. We have even seen how, in *Truth and Method*, he constructs an obviously normative hierarchy of three forms of understanding by which work in the human sciences is done. Of these three, the hermeneutical form of understanding is (to put it in formal terms) Gadamer's public normative choice in the human sciences and implies a public moral recommendation for practice. In both realms the judgment is that dialogue is good.

From these and other evidences, we are led to the conclusion that Gadamer's philosophy of human understanding cannot make its scientific point without also asserting an axiology: when understanding is defined by Gadamer as finding a 'common reality,' as the 'fusion of horizons,' these concepts are as closely related to the norm of solidarity as they are to the phenomenon of dialogue. In short, *how* we understand reality—by dialogue—has an essential bond with *what* is understandable about reality—such as, among other things, that solidarity is good. So, to engage in dialogue is good because it means the development of friendship. And every dialogue is the incarnation of some solidarity, of some friendship, and of at least one common reality. In fact, if we use 'dialogue' and 'solidarity' to refer to Gadamer's ontological and moral visions respectively, we cannot help but see an underlying circularity to these streams of his thought which blur all sharp distinctions between descriptive and prescriptive thinking, between how we understand and what we understand. Descriptively, dialogue is the 'condition' of solidarity; but prescriptively, solidarity is also the 'condition' of dialogue.

The project of a practical philosophy informed by the hermeneutical perspective can be enriched by developing this basic and almost obvious implication of Gadamer's work. It certainly helps to establish—with Gadamer's help and without it—the fundamental significance of his hermeneutical philosophy for reflection on moral practice, and this deserves future inquiry.

---

21 Ingram, 49.

Toward Co-ordination Between Hermeneutics and Traditions

The path just surveyed seeks to exploit the relevance of hermeneutical philosophy for practical philosophy beyond the point to which it is developed by Gadamer. But even in this case, it would still leave the project of a reasonably coherent and complete practical philosophy unfinished. No matter how much can be found in philosophical hermeneutics of relevance to the tasks of practical philosophy, it will be short of what is necessary: philosophical hermeneutics is not a substitute for practical philosophy, and its purpose is not centered on practice.

The materials needed to fill this gap in the practical philosophy which Gadamer bequeaths to his readers are available in many sources, in sedimented traditions and living minds. We would seek some coordination—perhaps of alliance or even congruence, but at least of compatibility—between these sources and the principles of hermeneutics. This is a second path for the development of practical philosophy which can be pursued independent of, or in tandem with, the first path. It is also a path which deserves exploration beyond the concerns of the present inquiry.

We also add that, in our view, such a project requires a new consideration of what functions as a "source." Fields like theology and mythology, natural science and cosmology (both mythical and scientific), must not be shunned but rather allowed into philosophical reflection, and likewise, philosophy should be invited—even compelled—to consider their inquiries as its own. The story-teller or theologian or cosmologist—each of these love wisdom differently, but there is no reason to presume it is less than the philosopher. From this rehabilitated perspective, practical philosophy can again portray the whole cosmos for us, and can thereby nurture our confidence in the possibility of good practice.

The etymology of the terms 'hermeneutics' and 'theory' can illuminate this possibility. 'Hermeneutics' is derived from the Greek verb *hermeneuein*, which appears to have an ancient link with Hermes, the Greek god who had the "function of transmitting what is beyond human understanding into a form that human intelligence can grasp." This is an arresting thought. A further provocation lies in the fact that the priest who mediated the utterances of the Delphic oracle was known as a *hermeios*.[22] Then the Greek origin of our word 'theory' suggests an important relationship between hermeneutics and theory. Nicholas Lobkowicz explains that *theoria* was "the official title of the group of state-ambas-

---

[22] Richard Palmer, *Hermeneutics*, 12, 13.

sadors which a city-state delegated to the sacred festivals of another city-state." The historical appearance of this term, then, is associated with human bridge-making in social, political and religious dimensions. "As such sacred festivals were usually connected with sports and games, *theoros* [the singular form of *theoria*] simply came to mean spectator," and even to describe any person who traveled far, like these ambassadors, to learn of the customs of foreign lands. But behind this usage was an even earlier reference. *Theoros* "originally referred to the envoy sent to consult an oracle."[23]

One who theorizes is still today a kind of spectator, but the original object of encounter was more queried than observed, more solicited than analyzed, and was in fact considered holy: the oracle by which humans had a doorway to what is divine. In that original dialogue with the sacred, the priest mediated the oracle's wisdom to the envoy. Thus a particular kind of experience connected the Greek words from which we, centuries later, derive our terms 'theory' and 'hermeneutics.' That experience, contrary to both scientific reasoning and to misunderstandings of what hermeneutics is about, suggests an idea which should invert the way we approach reasoning about human life and practice: instead of hermeneutics relying on theory, theory must wait upon hermeneutics, as the envoy (*theoros*) waited upon the oracular priest (*hermeios*).

There may be no clear separation between the two paths we have briefly considered, in fact, each path encourages exploration of the other path. The second path, of co-ordination between the dialogical ontology and external axiologies, should not leave the impression that philosophical hermeneutics can or should be reduced to a storehouse of value-neutral abstract claims—exactly the devolution that would betray Gadamer's efforts and which he himself probably would strongly oppose. Rather, as we saw Gadamer emphasize in Chapter Two, there is no firm division between historical science and tradition itself, and likewise here, Gadamer's hermeneutical philosophy already embodies the values of an age, and of its author as well. The first path, of bringing this inherent but hidden axiology in hermeneutical philosophy to light, looks inward for unexplored resources. There is obviously a horizon to this path as well, beyond which it becomes silly to try to derive norms from Gadamer's work which it cannot possibly support and for which we must look elsewhere. But Gadamer's noisy silence in opposition to

---

[23] Nicholas Lobkowicz, *Theory and Practice: History of a Concept from Aristotle to Marx* (Notre Dame, IN: University of Notre Dame Press, 1967), 6. Transliteration of Greek added.

this path does not change either what can be found in his work, nor the need of practical philosophy for such resources.

CORRECTING THE CORRECTIVE

Why does Gadamer stop? This is the second question posed at the outset of this chapter. Why does he not proceed in the directions we have just proposed, instead of remaining silent about much that is critically important to the viability of practical philosophy today? What unspoken reasons lie behind this silence? What identifies and explains the limit at which he ceases to elaborate on his philosophy of practice? Having completed most of our inquiry, the reluctance of our primary source to continue in the direction in which his own philosophy points is certainly a matter of concern. The fairest procedure is to listen: to consider the possible reasons for his silence together with the objections he might have to our notion that practical philosophy has a participatory function as well as a reflective function. But the merits of pursuing this question go beyond ad hominem inquiry. By identifying and addressing the choices which contribute to Gadamer's silence, we intend to show why no one needs to remain as silent as Gadamer—indeed, why we must not.

Some reasons for Gadamer's silence can be easily identified and as easily dealt with. It is clear, first of all, that he wishes to engage many issues, as the large scope of his work demonstrates. Time devoted to some areas lessens attention to others, and, as indicated earlier, in itself this is not a choice which can be criticized. But if he understood that such a philosophy required more, and knew that he was stopping short of that, it would not be unreasonable to expect him to so indicate, and to explain where and why he stops.

Of course, Gadamer does state certain intentions concerning how he thinks his philosophy should be perceived. We have seen various indications that he does not wish to enter—nor to be judged as entering—upon any level of prescriptive or even simply normative judgment in any realm, whether methodological, moral or otherwise. Perhaps clearest of all is Gadamer's statement, referring to philosophical hermeneutics, that, "it is not my intention to make prescriptions for the sciences or the conduct of life, but to try to correct false thinking about what they are." [24] Of

---

[24]   *TM*, xxiii [xiii]. If the reference to "conduct of life" is not clear enough to include practical philosophy, then let it be recalled that, given the deep affinities Gadamer sees between hermeneutical philosophy and practical philosophy, his intentions for

course, what may appear here to be an obvious lacunae and a dubious silence may be a cause for satisfaction in his own eyes. If indeed the delimitation just quoted reflected the pledge he made to himself and his readers, then our conclusion does indeed mean that even though we may be able to find moral judgments here and there, Gadamer by and large adheres to that pledge.[25]

But public intent is one thing; his unspoken assumptions and unannounced convictions are another. In Gadamer's case, these are so evident that a number of things can be surmised about his reasoning, despite his protestations to the contrary. Such a reading of Gadamer suggests that when he chooses to be silent, in addition to such reasons as we have already proposed, he is also *avoiding* the participatory function, and that he is in fact *reluctant* to say more. And this in turn suggests that behind his silence there are other reasons, philosophical or otherwise, consciously intended or not, which wait to be illuminated. Indeed, these reasons are readily apparent and draw attention more than once or twice.

Of course, it is impossible to identify such unspoken intentions with certainty. Nor is it profitable to the larger ends of shaping a viable practical philosophy to be immoderately absorbed in debating the exact nature or degree of his apparent reluctance. But there is benefit in demonstrating what seems to be the origins of Gadamer's reluctance. One benefit is to show those who, like Gadamer himself, seem to see some good purpose in this silence, why it is incompatible with the hermeneutical principles they have embraced, or at least why we can speak up without compromising those principles. As a consequence, there will also be benefit to those who wish to further build a hermeneutical philosophy of practice, and want to know and justify why they need not be as silent as Gadamer and still remain his beneficiaries. They want to know how to avoid Gadamer's errors as well as those of Strauss and Habermas.

The reason for Gadamer's silence—at least a primary reason for it—is that he succumbs to a temptation. Of course, he aims to make certain temptations evident to his readers, and his critique of Habermas and Strauss illustrate this concern. But in the course of protecting us from such temptations, and also in order to protect us, he yields to a different temptation.

---

his philosophy of hermeneutics could be expected to largely apply to his philosophy of practice as well. See also *TM*, xxiv [xvi].

[25] Of course, his most sustained failure to observe this pledge involves his frequent criticisms of the technocratic tendencies of contemporary practice: it is too bad that this outburst mars his promised silence.

Richard Bernstein has written an essay on "Hermeneutics and its Anxieties."[26] The anxiety of which he speaks—aroused by the prospect of unmitigated relativity—is certainly evident in many writers, among them both supporters and critics of Gadamer, and is demonstrated in the challenges posed by Habermas and Strauss, and by our own concern to preserve a hermeneutical philosophy of practice from relativism.[27] As we have seen, Gadamer is also concerned to embrace the historicity implied by relativity without admitting that this opens the gates of moral arbitrariness. But Gadamer's most basic anxiety is actually not of this kind. He is not most worried about whether the conclusions of hermeneutics lead to the dilution of moral abstractions, or accelerate the seeming slide into amorality of this age of tolerance and plurality. Gadamer's anxiety is more like that of the Protestant—an anxiety about human arrogance, and most of all one's own arrogance—but here embodied in anxiety about the desire or need to observe certain absolute standards. This fact about Gadamer's perspective is as revealing about him (and perhaps about his own religious ethos) as it is disturbing that it has slipped past recognition by others. The primary task here, though, is evaluation of the measures Gadamer takes, driven by this anxiety, to avoid such dangers: do they warrant his silence?

The hermeneutical perspective, as we have seen Gadamer present it, is concerned to avoid—and rightly so—two dangerous temptations, which often appear in tandem: first, the danger of foundationalism—of claiming to have absolute knowledge, when in fact finite beings cannot possess such knowledge; and second, the danger of technology—of attempting to bring about good by displacing *phronesis* and technicizing practice, when in fact good depends on the judgment of *phronesis* and cannot be produced. Both of these dangers are essentially temptations to

---

[26]   Richard Bernstein, "Hermeneutics and its Anxieties," 58-70, in *Hermeneutics and the Tradition*, ed. Daniel Bahlstrom (Proceedings of the American Catholic Philosophical Association, vol. 62), Washington, DC: ACPA, 1988.

[27]   Hermeneutics does mean giving up a craving for absolutes, a craving for firm foundations, for clearly specified decision procedures by which we can adjuticate among conflicting interpretations. There is no guarantee that there will be a convergence of interpretations. The thrust of hermeneutics is to show us that reasonableness is more open, fallible, flexible and historically conditioned than many philosophers have led us to believe. But the specter that haunts us is that once we take the hermeneutical turn we are led down the slippery slope where 'anything goes.' (Bernstein, "Hermeneutics and its Anxieties," 67.)

arrogance—the arrogance of attempting to transcend our historicity, on the one hand, by claims for absolute knowledge, or on the other hand, by claims for the ability to produce good. Gadamer also raises a concern, evidently more on his own than in the name of hermeneutics, about appeals within society to the special authority of the philosopher, appeals which he considers dangerous to both philosophers and the rest of society because, once again, they tempt to arrogance, here in social relationships. By exploring these three topics in the concluding part of this chapter, we can uncover the reasons for Gadamer's silence: the implied warnings which appear in his texts, and still deeper, the unspoken anxiety which is more than sufficient to account for his defensive measures and therefore appears to lie behind his silence as well.

The following warning could well apply to all three of the dangers which concern us here:

> The Delphic demand "Know thyself" meant "Know that you are a man and not a god." It holds true as well for human beings in the age of the sciences, for it stands as a warning before all illusions of mastery and domination. Self-knowledge alone is capable of saving a freedom threatened not only by all rulers but much more by the domination and dependence that issue from everything we think we control.[28]

Would the participatory function of practical philosophy, as we have described it, necessarily lead to the arrogance which Gadamer is particularly concerned to avoid—and which gives rise to the above statement?[29] As we have seen in several instances, Gadamer seems to acknowledge that practical philosophy entails this function. It is therefore reasonable to think that concern about the dangerous consequences of such an effort may have been a major factor in Gadamer's relative silence about the participatory content of his own practical philosophy. But there are certain philosophical positions and tendencies in Gadamer's texts which appear to be protective measures against the above mentioned dangers of arrogance; they also happen to preclude the adequate articulation of a participatory content. In fact, these protective measures are the most apparent evidence that Gadamer harbors the implied and unspoken concerns postulated above. We will identify these measures and then

---

[28] *RAS*, 150.

[29] The participatory function, to reiterate, includes articulating assertions about moral universals and about the ultimate significance and source of those universals, and it involves theoretically and hypothetically applying such assertions to practical issues.

respond to them, in order to see whether they persuade us to rethink our defense of the participatory function. If engaging in reflection for this purpose leads to making these mistakes, it would seem best not to address the participatory function, or to do so as little as possible, or in the most general terms possible. If Gadamer's concern is unwarranted, then his silence reveals more about his anxiety than about our foolhardiness.

ON THE DANGER OF FOUNDATIONALISM

Gadamer's concern about the danger of foundationalism is perhaps most deeply expressed in his commitment to phenomenology, or to the "philosophophical methodology" which he associates with "the problem of phenomenological immanence."[30] This phenomenologism—for we may contrast it to a commitment to foundationalism—at least means that philosophical hermeneutics makes no claims for foundational knowledge about ultimate reality, truth, or good. Moreover, Gadamer makes clear that, in his view, no philosophy can legitimately make such claims:

> This fundamental methodical approach [of phenomenologism] avoids implying any metaphysical conclusions . . . I have recorded my acceptance of Kant's conclusions in the *Critique of Pure Reason*: I regard statements that proceed by wholly dialectical means from the finite to the infinite, from human experience to what exists in itself, from the temporal to the eternal, as doing no more than setting limits, and am convinced that philosophy can derive no actual knowledge from them.[31]

A similar point is made to Habermas when Gadamer says that "with [the] area of what lies outside the realm of human understanding and human understandings (our world) [*sic*] hermeneutics is not concerned."[32] Strauss is also addressed along these lines, as observed in Chapter Three, when Gadamer says that "philosophy must learn to do without the idea of an infinite intellect."[33]

In keeping with this phenomenological perspective, Gadamer asserts that his philosophy, including its dialogical ontology, does not make claims about the meaning or content of all reality or ultimate reality, as

---

30   *TM*, xxiv [xxiii, xxiv].
31   *TM*, xxxvi [xxiv].
32   "On the Scope and Function," 87.
33   "Correspondence Concerning *Wahrheit und Methode*," 10.

ontology is conventionally understood to be engaged in doing. Rather, philosophical hermeneutics is intended to be a more limited claim about how we understand the realities of the world which *are* intelligible to humans. That is, Gadamer is making a phenomenological claim that language has ontological significance as the horizon or scope of humanly intelligible reality. Consequently, he does not provide, and in his view does not need to provide, any ultimate or transcendental justification for his ontology or for those normative assertions which he does make (eg., for the moral value of solidarity, or for his confidence about the possibility of true understanding and good practice), because they are part of his phenomenological and interpretive report on the immanent meaning of the world around him. The question of whether we can have more knowledge about or more confidence in moral truth than by phenomenological means is moot for Gadamer since it asks about what is impossible for finite beings to have.[34] Gadamer views this phenomenological approach as one which should also characterize the enterprise of practical philosophy:

> Aristotle asks, "What is the principle of moral philosophy?" and he answers, "Well, the principle is *that*,—the thatness." It means, not deduction but real givenness, not of brute facts but of the interpreted world . . . And so my thesis is: exactly because we give up a special idea of foundation in principle, we become better phenomenologists, closer to the real givenness, and we are more aware of the reciprocity between our conceptual efforts and the concrete in life experiences.[35]

Thus phenomenologism is the means by which Gadamer seeks to orient his work away from the danger of foundationalism.

---

[34] He implies that the absence of ultimate grounds for his own philosophical claims does not worry him:
> Hence present investigations do not fulfill the demand for a reflexive self-grounding [of philosophical hermeneutics] made from the viewpoint of the speculatively transcendental philosophy of Fichte, Hegel, and Husserl. But is the dialogue with the whole of our philosophical tradition—a dialogue in which we stand and which as philosophers, we are—purposeless? Does what has always supported us need to be grounded? (*TM*, xxxvii [xxiv].)

In effect, Gadamer is arguing that every philosophy that tries to ground itself in a foundationalist theory is only using its own claims to justify itself, and that in fact philosophy can stand without such attempts. Consequently, his own work is presented without such a foundationalism.

[35] "Hermeneutics of Suspicion," 65.

The precise nature and ramifications of this phenomenologism, however, are not clear. On the one hand, it appears to release practical philosophy from the burden of attempting to guarantee its assertions, and opens it to 'what is' and perhaps to greater recognition of the reciprocity of theory and practice. On the other hand, as preceding quotations suggest, his phenomenologism seems to mean that it would be inconsistent for practical philosophy to assert moral universals or claims about ultimate reality. This doctrine of uncertainty is also supported by other statements about axiology. "The idea of the good lies beyond the scope of any science. This is very clear in Plato. We cannot conceptualize the idea of the good."[36] In fact, practical philosophy "makes no determinative use of arguments of a cosmological, ontological, or metaphysical sort for practical problems."[37]

It appears likely that those who, with Gadamer, affirm this phenomenologism would be concerned that the notion of a participatory function would necessarily lead practical philosophy to the arrogance of foundationalism. How, then, shall we respond to such an objection? We have already established that we share the view that an 'infinite intellect' is an illusion, and it is on this basis that we have built the conclusions of previous chapters; our concept of the participatory function is not intended as a means for reintroducing claims for foundational knowledge. However, we do not think that this non-foundationalism precludes assertions about moral universals, or about the ultimate ground of such universals, even though Gadamer's phenomenologism does not appear to allow this.[38] Instead of directing persons to never make universal or ultimate assertions, philosophical hermeneutics has the opposite effect, as discussed in Chapter Four: it takes for granted that we always and inevitably make such claims, yet it does not presuppose that doing so is to succumb to the arrogance of a self-deluding foundationalism. The problem of philosophical arrogance should be dealt with not by silence about moral universals and ultimate realities, but by nurturing a hermeneutical consciousness about our assertions.

---

[36] "Gadamer on Strauss," 12. Also in *TM*: "There can be no anterior certainty concerning what the good life is directed towards as a whole." (321 [287].)

[37] *RAS*, 117.

[38] And if it does not, so much the worse for Gadamer's phenomenology—although perhaps our efforts will still conform to another definition of phenomenology. The immediate point here is to follow how, in fact, humans understand the world and themselves.

When practical philosophy is guided only by a phenomenologism which prohibits such moral assertions, practical philosophy becomes separated from tradition—which, as we have repeatedly observed, is exactly Gadamer's tendency—and the result is a content lacking a clear axis, a content which will not adequately serve the practical tasks to which it is directed. Moreover, if phenomenologism means an inhibiting cautiousness in making assertions about things around us, about the content of the good 'here and now,' then it has not really liberated us from the prejudices or demands of foundationalism. No doubt Gadamer thinks his caution is justified by a desire to protect philosophy from being cheapened by allowing abusers to claim that their philosophizing can solve all sorts of problems. But in his zeal, he is depriving philosophy of sustenance it needs, especially today. Should we allow the scope and function of philosophy to be determined by a definition which is designed primarily to exclude the abusers of philosophy? Will we not then be distracted from the main beacon of philosophy, which will shine on, even without our efforts to keep it burning?

As this suggests, Gadamer may be guided, more than he realizes, by the assumptions of our age about what constitutes philosophy, as this appears, for example, in the relationship between 'theory' and 'tradition' (or, similarly, between 'theory' and 'ethos'). In Chapter Two, we saw how hermeneutical philosophy asserts that there is no dichotomy but rather a continuity between human science and tradition,[39] which should apply analogously to the relationship between theory and tradition. Yet Gadamer often writes as though he thought this relationship was still a dichotomy. He seems to labor, perhaps unconsciously but quite ironically, under the notion that theory and tradition are two clearly distinct and even autonomous realms of human activity. This was particularly evident in "Über die Möglichkeit einer philosophischen Ethik," as discussed in Chapter Five. Gadamer speaks there of how the traditions of our social life define us, but implies that those norms never appear to us as theory or in a theory, a form of human activity apparently of a higher order. Conversely, the theorizing he associates with philosophical ethics seems to be only a kind of abstracted thought which organizes formal observations, but which has no evident connection with the actual substance of tradition. In short, only theory is associated with philosophy, presumably leaving tradition and ethos to be judged 'less rational' or 'pre-rational.' Gadamer seems to doubt that it is the job of philosophy

---

[39] *TM*, 281-282, 282-283 [250, 251].

to do what we ask of it, or thinks that such things should be 'left' to tradition.

But if practical philosophy is to fulfill its purpose, this continuity between theory and tradition must become evident in its own assertions. And since traditions make universal and ultimate assertions—and do so in what we call mythical, poetical, religious and cosmological languages—practical philosophy must be able to speak in the same languages.

Shouldn't the reasoning of practical philosophy be in dialogue with the reasoning of *phronesis* and with all reasoning that bears on practice? The reasoning of practical philosophy is clearly different from the reasoning of practice itself; but the question is whether the former should exclude all substance and even semblance of the latter (which seems to be the character of what Gadamer actually does as distinct from what he declares), or whether practical philosophy should acknowledge the presence of practical reasoning, as we think, and even welcome it.[40] Would we then be expecting of practical philosophy what should only be expected of myth, or of belief, or of tradition? This seems to be Gadamer's view, but it also exposes a prejudice of his own: that communicable reason reaches its limit at the border of philosophy, and that most of the reasoning which is relevant to practice is incommunicable and thus beyond the scope of practical philosophy. Such a view cannot be long sustained: it leads ultimately to the demise of practical philosophy itself, since it increasingly relegates practical philosophy to the sidelines of actual practical problems, where it becomes something other than an aid to practical judgment. Moreover, such a view implies that in the regions of myth or cosmology which lie beyond this border, reason does not exist.

Gadamer hardly seems the philosopher to embrace these views, yet this is the direction in which his writings lean. His implied view of reason in regard to this matter is characteristically evident in the closing passage of a 1971 address. This is another work still untranslated and little noted in secondary literature, which is entitled, "Das ontologische Problem des Wertes" ("The Ontological Problem of Value").[41] His overriding aim seems once more to be critical, corrective and defensive. Practical philosophy, he says, always must presuppose the existence of a

---

[40]   And if Gadamer agrees with us, then why is there so little evidence that he too welcomes these things?

[41]   *Human Sciences and the Problem of Values*, ed. K. Kuypers (The Hague: M. Nijhoff, 1972), 17-31.

concrete and specific ethic, but from which it also must always be distinguished. Practical philosophy sets borders within which that ethic is visible, without, however, being able to provide, on demand as it were, the actual content of the ethic. Such a demand, he says, is a "fantastical illusion of theoretical reason." Gadamer goes on:

> Therein [i.e., in the impossibility of this fantasy] lies the limitation but also the legitimization of all "practical philosophy," which turns away all hope of lifting our eyes to a free glimpse past the completely overcast sky of value; all such supposed exploration is exposed as ethical self-deception, which does not widen an ethos that is too narrow, but disavows and corrupts every ethos.[42]

The hallmark of Gadamer's view on this matter, as recapitulated by this passage, seems to be one of keeping the substance of ethics, and the reasoning of practice itself, at arm's length—not quite rejected but not embraced either. There is more than a subtle difference between this posture and one which permits the tension between theory and practice to be a dynamic one which is allowed to take effect on practical philosophy.

To put the matter differently, Gadamer makes no room in his own philosophizing for the very kind of knowledge and reasoning on which he says we ought to rely. He excludes it, apparently because it cannot be smoothed into the shape of what he takes philosophy—or reason—to be. He keeps telling us that there is a practical knowledge different from theoretical knowledge. But is it therefore a knowledge which cannot ever be admitted to the table where philosophizing occurs? In Gadamer's eyes, it may not even reach outside of the specific situation of judgment in which such knowledge is apprehended as insight, to use Gadamer's important term. But we can quite legitimately remind Gadamer of what he himself asserts: that each philosopher, including Gadamer, is already in a specific situation! What emerges from each of them as theory is, from the viewpoint of hermeneutical consciousness, the fruit of—precisely—insight. Theorizing *is* a practice.[43]

---

42 "Ontologische Problem," 31. Both quoted passages from this essay are translated by Amy Brougher and revised by this writer.

43 In the following, Gadamer explicitly acknowledges what he more often seems unable to recall: "'*Theoria*' is not simply in opposition to *praxis* but rather is itself a highest *praxis*, a highest way of being human" ("Über die Möglichkeit," 179.). The same point is made in a less abstract way here: "My thesis is that theory is also an original anthropological datum, just like practical and political power . . . I am convinced that human society exists only because and as long as such a balance continues to exist." ("Science and the Public," 166.)

Natural Law

Our rebuttal to Gadamer is further supported, although uninten-
tionally, by Gadamer himself when we pause to consider that, in his
view, the articulation of natural laws is compatible with his non-foundationalism. It follows that such an articulation could be engaged in as part
of the participatory function.[44] Gadamer observes that Aristotle has long
been understood to contrast the transcendence of natural law to the
fallibility and changeability of positive, or humanly created law. But he
argues that "the true depth of [Aristotle's] insight has been missed." He
understands Aristotle to have made a further distinction between "the
idea of an absolutely unchangeable law," which pertains to divinity (on
which Gadamer does not add any comment in *Truth and Method*), and
natural law. Natural law refers to "things which do not admit of regula-
tion simply by human convention" but which are still not the same as
divine law. Here, "the 'nature of the thing' constantly asserts itself." That
is, every historical reality or being is not mute but keeps announcing
what its reality is and what it means. "Thus it is quite legitimate to call
such things 'natural law.'"[45]

But whatever asserts itself is also something which changes. "In that
the nature of the thing still allows some room for play, natural law is still
changeable." The most illuminating example which Gadamer draws
from Aristotle concerns the nature of the good political state: Aristotle
"quite clearly explains that the best state 'is everywhere one and the
same', but it is the same in a different way that 'fire burns everywhere in
the same way, whether in Greece or in Persia.'"[46]

Thus in Gadamer's reading of Aristotle, natural laws are not "mere
conventions" like positive laws; but it is also true that "they are not
norms that are to be found in the stars, nor do they have an unchanging
place in a natural moral universe, so that all that would be necessary
would be to perceive them." Rather, natural laws "really do correspond
to the nature of the thing—except that the latter is always itself determined in each case by the use the moral consciousness makes of them."[47]
For Gadamer, then, natural law is something which is both the universal
and the particular. Natural law must be discerned in the midst of the

---

[44] The relevant passage in *TM* is 318-320 [284-286]; the same topic is also discussed
in "Hermeneutics and Historicism," in *TM*, 518-520 [471-471], and in "The Problem of
Historical Consciousness," 34-35.
[45] *TM*, 319 [285].
[46] *TM*, 319, 320 [285].
[47] *TM*, 320 [285, 286].

historicity and changeableness of life and of moral judgment itself. This was implied already in Gadamer's statement, quoted in Chapter Five, that "there is a right according to nature," which is said in such a way (*"Es gibt ein von Natur Rechtes."*) as to make clear that he is not invoking the conventional meaning of natural law or natural right (*Naturrecht*) as an unchanging, absolute moral law.[48]

Rather, Gadamer means that the conditionality of the human ethos is compatible with the unconditionality of the moral. Aristotle's view of natural law, therefore, "can help us avoid falling into an apotheosis of nature, naturalness and natural law," and instead lead us to the real nature of the object. More generally, Aristotle's ethics provides "a critique of the abstract and universal that . . . has become essential for the hermeneutical situation after the rise of historical consciousness."[49]

Practical philosophy, then, should not give undue authority to natural laws since they are in fact less abstract and more changeable than the popular view would have us believe, yet neither can they be created like conventional or positive law. In this sense, Gadamer's interpretation of natural law serves one of the purposes of the reflective function of practical philosophy: it restrains us from giving too much authority to any abstract theoretical claim. However, a different implication of his discussion needs to be elucidated as well: if natural laws are not reified abstractions nor divine laws, then by implication they are more accessible to moral reasoning than is generally thought to be true. In this case, the articulation and application of natural laws are activities well within the scope of possibility for historical beings, and are not necessarily synonymous with foundationalism. Furthermore, if one so wished, the participatory function could still be addressed as a phenomenological task insofar as natural law and phenomenologism both are focussed on allowing the 'nature of the thing' to show itself.

ON THE DANGER OF TECHNOLOGISM

The second danger of which Gadamer is particularly aware, is that of thinking we can control and produce good by displacing *phronesis* and technicizing practice. Gadamer's emphasis on the distinction between practical philosophy and *phronesis* is the most significant step by which

---

48 "Über die Möglichkeit," 191.
49 "Hermeneutics and Historicism," in *TM*, 540-541 [490].

he aims to prevent this danger. Efforts to technicize practice presuppose that theory can generate practical wisdom, but, as we have seen in Chapter Two, Gadamer makes it quite clear that "practical philosophy is not the virtue of practical reasonableness," that is, *phronesis*.[50] The meaning of good practice which Gadamer seeks to protect by affirming this distinction involves the nature of human judgment: "The knowledge that gives direction to action is essentially called for by concrete situations in which we are to choose the thing to be done; and no learned and mastered technique can spare us the task of deliberation and decision."[51]

From the perspective of Gadamer's efforts to distinguish *phronesis* from practical philosophy, the idea that practical philosophy includes a participatory dimension—of applying moral universals to concrete problems—might appear to be an idea that would inevitably reduce practical philosophy to a technology designed to displace *phronesis*. In reply to this concern, we first reiterate the scope of our agreement with Gadamer: the contents of practical philosophy cannot substitute for the judgments we must make through *phronesis*, and our concept of the participatory function is not intended to displace *phronesis* or to technicize practice. Indeed, the participatory content must always be something more imprecise than what actually occurs in deliberation. As Gadamer points out, all we can do is apply "more or less vague ideals or virtues and attitudes to the concrete demand of the situation."[52] But Gadamer takes this vagueness too literally; in fact he seems to insist on it. The distinction between practical philosophy and *phronesis*, even though it is designed to serve the same purpose, is usually drawn so sharply that it amounts to the imposition of a dichotomy between theoretical knowledge and practical knowledge.

For example, Gadamer says that the various fields of theoretical knowledge, including practical philosophy, contain "not the knowledge of a physician or a craftsman or a politician that is always to be found in application but knowledge about what may be said and taught in general about such knowledge."[53] If the two kinds of knowledge are so different, there is very little—perhaps even nothing—to be said about the knowledge which *phronesis* has. This seems to be an over-reactive response by Gadamer to avoid risking any possibility that *phronesis* will be displaced by a moral technology. This line of thinking by Gadamer is also implied

---

50   *RAS*, 117.
51   *RAS*, 92.
52   "Practical Philosophy as a Model," 82.
53   *RAS*, 116.

in the contrast—even the gap—which distinguishes the two following statements:

> The question of justice, the question of the perfect state, seem to spring from an elementary need of human existence. Nevertheless everything depends on the way in which this question is intended and asked, if it is to bring clarification.[54]

It is quite true—and also identifies a basic concern of Gadamer's reflective content—that the 'How' of deliberation and application is critical, and that its procedures cannot be codified. But so frequent is his recourse to this point, and so ubiquitous is his silence before what even he admits is the "elementary need" to affirm what justice or good polity means, that it leads one to think that his silence also reflects a reluctance to commit his practical philosophy to substantive assertions.

Such a dichotomy between practical philosophy and *phronesis* contradicts the reciprocity of theory and practice, the very claim which constitutes a basic pillar of philosophical hermeneutics. From the perspective of this reciprocity, there can be no artificial division between the practical knowledge of *phronesis* and the theoretical knowledge of practical philosophy; in fact, not only can we arrange for each to inform the other, but they also will do so beyond and before all our intentions. As the ultimacy of universals, such as natural laws, cannot be eliminated from practical judgment, neither can the concreteness of practical judgments be excised from practical philosophy. Thus, if practical philosophy is to fulfill its purpose in the light of its own reciprocal relationship with practice, its participatory function must be addressed. The risk that doing so will result in technicizing practice is real. But even Gadamer will acknowledge that this risk cannot be completely eliminated.[55] Moreover, the idea that engaging in the participatory function is the source of this risk is tantamount to asserting that the reciprocity of theory and practice, a phenomenon that characterizes all of human life according to philosophical hermeneutics, is itself the covert source of the temptation to technicize practice! The source of that temptation certainly deserves more attention

---

54 "Hermeneutics and Historicism," in *TM*, 541 [490].

55 With reference to textual interpretation, Gadamer says this: "Understanding, like action, always remains a risk and never leaves room for the simple application of a general knowledge of rules to the statements or texts to be understood . . . Understanding is an adventure and, like any other adventure, is dangerous." (*RAS*, 109-110.) If this applies to the understanding of texts, it must also apply to the understanding of practice, and especially to practicing good.

than Gadamer gives it or than we will be able to devote; but Gadamer would surely agree that the source of it lies elsewhere.

On occasion, Gadamer reinforces the difference between practical philosophy and *phronesis* in a way which also implies—in spite of himself, it seems—the very opposite, that is, that the participatory function is a reality despite any risks that may accompany it. In his essay on philosophical ethics, Gadamer states this: "Moral knowledge is not completed in the general concepts of bravery, justice, etc., but rather in concrete applications that define what is practicable here and now, in light of such knowledge."[56] Authentic ethical knowledge is said to be the knowledge which *phronesis* acquires or discovers in the practical moment, not the knowledge which a theory or philosophy has in the form of "general concepts." However, since Gadamer neither attends to them nor clearly affirms the positive role of "general concepts," his statement leaves the clear impression that there is really no use for them. But it is not true that we can do without such concepts, as his own phrasing suggests: the knowledge of *phronesis* appears "in light of" those "general concepts." Providing that illumination is precisely the purpose of the participatory function.

A recent defense of Gadamer's views in these matters by Ronald Beiner presents an opportunity to clarify our position by means of a critical response.[57] In doing so let us remember that the critique made of Gadamer in this part of our investigation is not merely an offensive, nor is it targeted only on Gadamer himself, or on supporters like Beiner. It is an opportunity to reconsider the best way to articulate practical philosophy. For Gadamer's errors are not the only problems at issue here; we want to identify and avoid the errors of the age to which Gadamer has been responding, an age swept away by an illusion of the salvific potential of its own reason. The point is not only to advance beyond Gadamer's limitations, but to do so with his adversary in mind as well, to correct his corrective so that the era of historical consciousness can be given a better compass bearing. Keeping this in mind, Beiner's important

---

56 "Über die Möglichkeit," 187.

57 Ronald Beiner, "Do We Need a Philosophical Ethics? Theory, Prudence, and the Primacy of *Ethos*," *Philosophical Forum* 20, no. 3 (Spring 1989): 230-243. Although certain critical problems with this essay are raised here, it also presents a number of important points which reinforce the positive impression created by our earlier discussion, in Chapter Four, of a passage in his book, *Political Judgment*.

essay also serves the useful function of illustrating how, in certain key respects, *not* to defend Gadamer.[58]

The main title of Beiner's essay, "Do We Need a Philosophical Ethics?," indicates that he has seized on a central question in our own inquiry, and to answer it he relies significantly on the essay by Gadamer which was examined in Chapter Five, "Über die Möglichkeit einer philosophischen Ethik." His objective opens up the heart of the problem left by Gadamer to those who follow in his steps: Beiner says that he seeks "a heightened awareness of the gap between a *theory* of practical reason and the concrete demands of practical reason itself." He hopes thereby "to defend neo-Aristotelian ethics [and presumably Gadamer's ethics in particular] against Habermas' charge that it rests solely upon, as he [Habermas] disparagingly terms it, substantive 'worldviews.'"[59] One wonders what worldview Beiner thinks Habermas found: the evidences which might be adduced for this—that Gadamer has blessed the status quo, or has embraced a diffuse and naive idealism, even if they were true—do not a worldview make. It is precisely the more specific elaboration of any worldview at all which we have unsuccessfully sought in Gadamer's work. To be sure, there is a worldview in Gadamer's work; it is simply so imbedded in and under his consciousness that in its present state it has little relevance for practical philosophy.

By contrast, Beiner's aim is to raise Gadamer's practical philosophy above any suspicion of containing a worldview by pointing precisely at the gap between theory and practice which (as we have also noted) is emphasized by Gadamer himself. This gap plays an important role in the answer which Beiner thinks Gadamer would give to his title question, and the answer which Beiner himself embraces:

> No, we do not need a 'new' philosophical ethics, for the ethics already to hand within the tradition are perfectly sufficient, and in any case, no 'new' ethics can serve to restore the *ethos* that animates ethical practice if indeed that *ethos* has dissipated within the life-practices of our society.[60]

---

[58] But even if we have misinterpreted Gadamer's silence, or have perceived more in that silence than was there, the effort expended here will not be in vain. For if in fact he basically agrees with the position we have taken, then our common purpose is that much clearer: a defense of the viability and relevance of a hermeneutical philosophy of practice for an age which is too mesmerized by a fantasy about its own powers—but also unable to discriminate which powers are genuinely given to it.

[59] Beiner, "Do We Need," 230.

[60] Beiner, "Do We Need," 233.

This defense of Gadamer requires some analysis of the interpretation on which it is based. Beiner is correct that in terms of communicable resources the contents of tradition are sufficient. But by his emphasis on whether a *new* ethic is necessary, Beiner indicates that disputation with Habermas is a primary motivation here. The focus of concern for us, however, and in our view the more basic problem, is that even an 'old' ethic is still lacking in Gadamer.[61] From the hermeneutical perspective, it is also quite true, as Beiner indicates, that no practical philosophy will, by itself, restore a diseased or corrupted ethos to health. But the question is inappropriate since, according to the reciprocity principle, it is a mistake to even inquire about the role of theory in terms of whether, or how, or how far it will restore a diseased ethos to health, without also considering the mutuality of theory and practice in the full range of their historical effects on each other. (Moreover, as we learned in Chapter Five, it is not clear that Gadamer would accept Beiner's implication that, in order to secure good practice, we must primarily or even solely depend on the existence of a healthy ethos or robust tradition to the exclusion of other factors.)

These problems in Beiner's interpretation of Gadamer draw attention to the prejudices on which they appear to be based. It is obvious that Beiner intends to closely follow Gadamer in his interpretation of the relationship between theory and practice. In Gadamer's essay on philosophical ethics, this relationship is posed in the form of a perennial dilemma: in order to philosophize about ethical norms, one must (in Beiner's words) both "distantiate oneself from them" and "participate in them." Beiner characterizes Gadamer's view of this dilemma in terms of a "tense relation" and "imperishable tension" between theory and practice, and says "Gadamer's constant emphasis upon Aristotelian *phronesis* is meant to bring to mind a polemical opposition between abstract theory and concrete prudence."[62] (With this interpretation of Gadamer, Beiner supports our own sense that Gadamer often makes a radical distinction and separation between theory and practice. Moreover, if that distinction is primarily a polemical one, we still ought to be able to recognize his

---

61   We have seen that Gadamer is keenly aware of how deeply tradition, or ethos, needs our active affirmation and preservation as well as renewal. But if he thought this was so essential, would he not have taken greater pains to specify it? If we did not think Gadamer has unspoken or even forgotten reasons for his silence, we might assume, as suggested earlier, that Gadamer feels tradition is beyond the scope of human reasoning.

62   Beiner, "Do We Need," 234, 236.

real view; but since there seems to be little evidence of anything else, perhaps his polemical view is the only one he has.) These phrases of Beiner's indicate that he confuses the tension between theory and practice with an eternal antinomy. In Gadamer's case, such a mistake is precluded by his vagueness; for Beiner to rely so much on this "tension" and "opposition" is due to an over-zealous reading of Gadamer. Most important, Beiner does not see how his interpretation is simply inconsistent with the reciprocity principle on which Gadamer's own hermeneutical ontology is based, but of which Beiner seems unaware.

Perhaps it is the challenge posed by Habermas which leads Beiner to believe that practical philosophy needs a 'stiffer' backbone, not, of course, by making compromises with Habermas, but rather by adhering—more singlemindedly than Gadamer does—to what seems, to the modern outlook, the 'weakest' part of Gadamer's position: the claim that practice cannot be determined through the technical application of a theory. The challenge posed by Habermas, according to Beiner, is the view that *phronesis* is simply too unreliable when, as in our age, the ethos in which it finds nurture is itself severely damaged. But consider how Beiner responds to the Habermasian challenge in the closing pages of his essay. He chooses a passage from Gadamer's letter to Bernstein, which we also quoted in Chapter Five, as "Gadamer's clearest and most forceful answer to this challenge."[63]

That letter is indeed one of Gadamer's better responses to the Habermasian viewpoint, although as our survey has shown, that is not necessarily saying much. In fact, in that letter Gadamer does little to expand on his belief that the ethos on which we depend has not atrophied; he simply suggests that we could detect its absence if it truly had so decayed, and he calls upon the Platonic argument that there is honor of a kind even among thieves (implying that it derives from a larger sense of honor on which all of us can always rely). But this is irrelevant to Beiner's essay since the passage Beiner elects to quote contains none of this part of Gadamer's letter. Instead Beiner quotes Gadamer's response to whether *phronesis* or science is primary in practical philosophy. Of course, this is an important question and it bears on his debate with Habermas, although Beiner appears to use it primarily to reinforce his view of theory and practice as an opposition. But this is separate from— in fact, it is derivative of—his basic point in the letter: *phronesis* is never so bankrupt that it does not show its presence and its effects.

---

63 Beiner, "Do We Need," 240.

To be sure, Beiner means to remind us, correctly so and on Gadamer's behalf, that theory simply cannot solve the problems found in concrete practice. He is driven by a desire to defend the priority of *phronesis* over theory in practical judgment. "It is not any kind of moral skepticism that prompts [Gadamer's] doubts about the project of a philosophical ethics, but instead the worry that it will have very little to teach us at the real locus of our ethical experience."[64] In addition to this priority of *phronesis* in practice, the temptation to technicize theory is also at issue. In a footnote, Beiner translates and quotes the following from Gadamer's essay, "Über die Möglichkeit einer philosophischen Ethik:"

> The recipient of Aristotle's lectures on ethics must be immune to the peril of wanting to theorize simply in order to extricate himself from the demands of the situation. It seems to me that the enduring authority of Aristotle consists in his holding constantly in view this peril.[65]

But we must not let the Habermasian threat to these truths distract us from another truth. Beiner's defense of *phronesis* is half the story, the half which concerns practice; it must be complemented by the reminder that even practical wisdom is not omniscient, and can benefit from the fruits of theoretical reflection.[66] In fact, neither theory nor practice is 'omniscient': Gadamer reminds us, perhaps weakly but also undeniably, that we are beneficiaries of the possibility of *phronesis* which lies before and beyond any power of human choosing or causation, of thinking or acting.

In spite of these merits, Beiner's view of the relationship between theory and practice remains problematic, and this in turn is significantly due to Gadamer's vacillation on the matter. We never get a clear resolution as to whether Gadamer thinks the relationship between theory and

---

64 Beiner, "Do We Need," 236-237.

65 Beiner, "Do We Need," 242, note 20, translates and quotes Gadamer, "Über die Möglichkeit," 190.

66 In fact, Beiner's own point about the neglected Kantian aspect of Gadamer's ethical perspective supports this point. Beiner observes that Kant's search for universal principles is also evident in Gadamer, and in ways which Aristotle would not express or comprehend (Beiner, "Do We Need," 232-235 passim). The participatory function can be addressed in other ways than the Kantian, but if this too is to be regarded as legitimate theorizing in practical philosophy, it reinforces the legitimacy of the participatory function of practical philosophy. Actually, and ironically, we have gotten quite enough Kantian ethics from Gadamer in the form of the reflective content of his practical philosophy, but not enough Aristotelian ethics. This, in a nutshell, reiterates how the critique developed here is unlike the critique made by Habermas.

practice constitutes a dilemma or a reciprocity. He cannot alternate between these, as he often does, according to polemical utility; finally, it must be one or the other.

The point is not that we should try to drum up some grand theory (as though there were not ample evidence already that efforts at devising such a systematic ethic have failed too often in too many ways). As Gadamer often seems to be reminding us, what we seek is already right under our noses. But silence about what we find there does not necessarily make those realities any more obvious, nor does talk about them necessarily remove them from the concrete situation.

Gadamer's Prejudice Against Techne

Gadamer is so concerned to avoid the danger of technicizing practice that his treatment of *techne* is prejudiced. There is a latent and subtle bias against *techne* which appears often enough that it distorts his concept of practical philosophy and compromises the value of his critique of technologism in both of its forms, as methodologism in the human sciences and as technocracy in social practice. In effect, Gadamer makes technology his enemy, when in fact it is not knowledge about *techne* that is his worry, but technologism, the placing of undue trust in technology.

Several instances of this prejudice can be noted. At the beginning of his "Introduction" to *Truth and Method*, Gadamer speaks about the scope of hermeneutical phenomena in relation to technique as methodology:

> Even from its historical beginnings, the problem of hermeneutics goes beyond the limits of the concept of method as set by modern science. The understanding and the interpretation of texts is not merely a concern of science, but obviously belongs to human experience of the world in general . . . It [i.e., hermeneutics] is not concerned primarily with amassing verified knowledge such as would satisfy the methodological ideal of science—yet it too is concerned with knowledge and with truth.[67]

If, as Gadamer says here, the scope of hermeneutics goes beyond the scientific idea of method or technique, it must also include it. Technique is therefore a legitimate topic of a truly universal hermeneutical philosophy. It is true, as Gadamer says in the passage deleted from the above quotation, that "the hermeneutic phenomenon is basically not a problem of method at all;" but it is incorrect to infer from this—as Gadamer does in the same passage—that "it is not concerned with a method of under-

---

[67] *TM*, xxi [xi].

standing." For even if the focus of hermeneutics is not method, it is "concerned, here too, with knowledge and with truth."[68]

Signs of a bias against *techne* are also present in Gadamer's critique of the technocratic orientation of modern practice. As Richard Bernstein among others have also noted, there is almost no consideration of whether or how *techne* has a constructive role to play in modern practice, nor of how the technocratic temptation is related to other causes of unwise, deleterious or evil practice.[69] Under such conditions it can be easy to generalize from a critique of technocracy to a suspicion that *techne* itself is an autonomous and evil power, and Gadamer's commentaries in various places either sustain or do not sufficiently remove the impression that he has done just this.

---

[68] Paul Ricoeur says something which parallels the point we are making here. While he does not speak of any latent bias against *techne* in Gadamer, he does present a critique of Gadamer in the context of historical research which has affinity with our critique in the context of practical philosophy:

> The way back from a philosophical hermeneutic toward the methodology of historical research . . . is the more arduous path to follow. Yet it is on this way that hermeneutics has to test its capacity to contribute to an authentic critique of historical method. Its task is not completed with the return to foundations [i.e., Gadamer's ontology?]; it must end in a renewed dialogue with historical research. Yet the return to foundations may even risk becoming an obstacle or an inhibition to the carrying out of this second part of the hermeneutical task. The danger, in effect, is that hermeneutics may conceive of itself as standing in a purely dichotomous relation with historical method. (Paul Ricoeur, "History and Hermeneutics," *Journal of Philosophy* 73 [November 1976]: 690.)

[69] As an example of particular significance, Gadamer does not attend to the relationship of *techne* to power, an issue that naturally arises from his comments on global order and disorder. Bernstein takes a nearly opposite view from Gadamer's and argues that "despite contemporary transformations of the meaning and scope of the practical and the technical, the point that we need to be aware of is this: the danger for contemporary *praxis* is not *techne*, but domination." (Bernstein, *Beyond Objectivism and Relativism*, 156.) It is true that Gadamer does not adequately acknowledge that domination is indeed a more immediate and evident practical problem than is technology. But this critique of technology can still be incorporated into an awareness which is absent in Bernstein, and in Habermas for that matter: in our age, domination very much occurs through technology in new and little understood ways and needs to be studied from this perspective. As Gadamer implicitly seeks to show, the contemporary technocratic uses of *techne* are so generally presumed by societies to be legitimate that they constitute an ideology which is at least as powerful, and perhaps more invisible, than political or economic ideologies.

For example, Gadamer emphatically states that something "we need to learn from the classics" is the "absolute distinction between a politike techne and politike phronesis." Gadamer says this against Strauss, adding that "Strauss does not in my opinion give this sufficient weight,"[70] though our inquiries in Chapter Five suggest he would equally direct this point at technocrats. There is certainly a difference between a political order based on *techne* and one based on *phronesis*; but to speak of an "absolute distinction" implies a fundamental incompatibility between *techne* and *phronesis* themselves, as though *techne* would somehow corrupt *phronesis*.

In our earlier discussion of "Notes on Planning for the Future," we observed how Gadamer barely grapples with concrete issues about global order and disorder. As if to address the concern that practical philosophy should assist us further, he says this: "To demand more, . . . to expect that the solution of the inclusive difficulties of international politics will be provided by some kind of brain-trust, would be ridiculous."[71] He also states that "one should certainly not expect that, by becoming cognizant of the factors which set limits to the technological dream of the present, we could or should be able to influence the measured advance of progress."[72] When the absence of practical judgments about global issues by Gadamer is placed next to these sorts of passive and even fatalistic statements, one is left with the impression that consideration of global issues is either irrelevant to practical philosophy or is impossible. Since both such conclusions would contradict his own affirmation—that as "the science of the good," practical philosophy "promotes that good itself,"[73]—a different attitude toward *techne* is required.[74]

In *Truth and Method*, Gadamer speaks of his intention that philosophical hermeneutics expose as false the modern faith in *techne* as a model for knowledge and for practice:

> It seems to me, however, that the onesidedness of hermeneutic universalism has the truth of a corrective. It enlightens the modern viewpoint

---

70 "Hermeneutics and Historicism," in *TM*, 541 [490].

71 "Notes on Planning for the Future," 588.

72 "Notes on Planning for the Future," 587.

73 *RAS*, 118.

74 Lest we be unfair to Gadamer, recall that in "Notes on Planning for the Future" he also describes our modern task as that of piloting our way through the tension between scientific knowledge and practical wisdom.

based on making, producing, and constructing concerning the necessary
conditions to which that viewpoint is subject.[75]

It is noteworthy that he acknowledges his approach to this topic to be
onesided. This onesidedness, as he says, is intended as a corrective to
past errors, a force equal in power but opposite in direction to the
onesided character of modern science.[76] But onesided correctives always
run the risk of going too far in the opposite direction. Gadamer's tactical
purpose and the risk it entails are both implied here: "In my work,
heightening the tension between truth and method had a polemical intent. Ultimately, as Descartes himself realized, it belongs to the special
structure of straightening something crooked that it needs to be bent in
the opposite direction."[77]

In what respects does Gadamer make a onesided emphasis? In addition to themes that may more easily come to mind, such as the role he
gives to tradition, this onesidedness should also be understood in terms
of his treatment of the principle of application, occasionally labeled as
concretization. In itself the principle is not onesided, and certainly has an
intrinsic significance prior to its use as a "corrective" to the modern obsession with the technicity of making and producing. But the preceding
examples demonstrate that Gadamer's treatment of the truth of concretization goes beyond either of these purposes: his prejudice for the
truth of concretization is improperly enmeshed with another prejudice,
one against *techne*, which more distorts than illuminates, more abhors
than disciplines, the nature and conditions of "making, producing and
constructing."

In contrast to Gadamer's treatment of *techne*, an adequate concept of
it should begin with an acknowledgement that it cannot be excised from
life but plays an indispensable role in practice and in theorizing: *techne* is
part of human historicity. Gadamer's work tends to ignore the signs
which point in this direction. His commentary on Aristotle's discussion
of *phronesis*, presented in Chapter Two, notes the three ways by which
Aristotle distinguishes the moral knowledge of *phronesis* from technical
knowledge. One of these, which Gadamer reiterates in other works since
Truth and Method, concerns the fact that while technical knowledge involves knowledge of the means in practice, moral knowledge is unique

---

[75]   *TM*, xxxvii–xxxviii [xxiv].

[76]   This view of modern science also appears in his "Afterword," in *TM*, 2nd ed.,
552.

[77]   "Afterword," in *TM*, 2nd ed., 555.

in that it includes knowledge of ends as well, and judges ends and means together.[78]

> *Praxis* is not restricted to the special area of technical craftsmanship. It is a universal form of human life which embraces, yet goes beyond[,] the technical choice of the best means for a pre-given end. Aristotle's concept of prudence includes, as a matter of fact, the concrete determination of the end . . . Prudence as practical deliberation upon and discovery of concrete decision is both the finding of the means and the concretization of the ends.[79]

But this coherence of means and ends indicates something which Gadamer appears not to notice: that *phronesis*, like *techne*, has knowledge about the means to practical good and therefore must be familiar with technical knowledge in order to discipline it toward reaching practical ends. In fact, while Gadamer frequently emphasizes how *phronesis* and *techne* are different, he neglects to give constructive attention to the responsibility of *phronesis* for knowing about and judging means, a task in which we must coordinate technical and moral knowledge, and their different forms of application.[80]

After grasping Gadamer's distinction between moral and technical application, it must still be pointed out that this is only the first step in how practical philosophy should investigate *techne*. Practical philosophy must still inquire about the proper role of technical application within and as an aspect of moral application.[81] The popular view that *techne* alone is sufficient for judging practical means is a prejudice which must be challenged, but there is less benefit in doing so by ignoring the pres-

---

78   *TM*, 317-324 [283-289]. See also *RAS*, 92.

79   "Hermeneutics and Social Science," 312-313.

80   For example, Gadamer points out that while there is an age-old tension between a teachable science of universal knowledge, such as *techne*, and the individual's knowledge of practical good, this tension was not always viewed as an antithesis. Aristotle, for example, saw "no problems in the relation between political science and political sense (*techne* in relation to *phronesis*)." Only under modern science has this tension become an antinomy, with the result that *techne* is no longer seen as a "knowledge which fills the gaps nature left for human skill." ("Notes on Planning for the Future," 581, 582.) So if *techne*, whether of a materially productive kind or a rational kind, can be in creative tension with moral and practical knowledge, then we should expect hermeneutical philosophy of practice to deal more directly with the technical elements of practice and of its own reasoning than Gadamer has in his practical philosophy.

81   Bernstein makes a similar point in his response to Gadamer: for us as well as for the Greeks, he says, "*techne* without *phronesis* is blind, *phronesis* without *techne* is empty." (Bernstein, *Beyond Objectivism and Relativism*, 161.)

ence of *techne* in every kind of practice. Perhaps if Gadamer had further explored his own point about the bond between ends and means he would have more readily acknowledged the ubiquity of *techne* in practice. What Gadamer regards as a necessary defense of authentic practice is in significant measure an unjustifiable fear of *techne*. If the problem is the danger posed by *techne*, the solution is not to immediately abort practical philosophy but to inquire more deeply into the nature of *techne* and identify the source and location of the danger.

In one place, Gadamer clearly—and uncharacteristically—affirms the role of "know-how," or *techne*, in practice. In fact, the following statement represents an important part of the concept of *techne* which a hermeneutical philosophy of practice should develop:

> One will find in all spheres of practical application of rules, and thus in what one calls in general "practice," that the more one "masters" his know-how the more one possesses freedom vis-a-vis this knowledge. Who "masters" his art needs to prove his superiority neither to himself nor to others. It is old Platonic wisdom that true know-how makes possible precisely the distanciation from it. It is thus the master runner who can also run "slowly" the best, the one who really knows who can also lie most effectively, etc. . . . It is this freedom vis-a-vis one's own know-how which in fact liberates for the perspective of authentic practice that [*sic*; "and" seems more appropriate than "that"] which transcends the competence of know-how— what Plato calls "the good," which determines our practical-political decisions.[82]

Such mastery of know-how means exactly the mastery of *techne* in practical judgment. By the same token, however, practical philosophy too should not evade *techne* but 'use' it in order to aid our discernment of practical good. The philosopher of practice, no less than the agent of practice, must be one who masters know-how as it pertains to a given task or issue, and therefore has freedom in relation to it, rather than avoiding that know-how for fear of being mastered by it.

Techne *Reconsidered*

The following thoughts on the nature of *techne* and its role in theory and in practice can address this gap in Gadamer's practical philosophy. For our purposes here we focus on what appear to be the two basic forms of technology: productive technology and rational technology. *Productive technology* means the knowledge of how to make things and make them

---

[82] "Theory, Technology, Practice," 551-552.

serve as means; such things include not only visible objects like tools or machines, but the socio-cultural patterns—the invisible 'machines'—that shape the behavior and organization of human efforts which are needed to achieve visible ends. Productive technology makes our physical environment, and is a necessary technology because of the physical finitude and limitations of each individual, of each society, and of the species as a whole. *Rational technology* means the knowledge of how to make concepts, arguments and systems of thought, and how to use them as tools. It makes our intellectual environment, and is a necessary technology because of the finitude of every person's reasoning about the world and its meaning.

Obviously these forms of technology often (perhaps necessarily) occur together, and contribute to what we could call social or political technology (in a descriptive, non-pejorative sense). But productive and rational technology can be distinguished by the ways in which they correspond to practice and theory. The finitude which compels philosophers, for example, to use rational techniques is of a different kind than the finitude which compels practical agents to use productive techniques. Rational technology helps us to communicate and share ideas about what is unchangeably true, while productive technology helps us to generate effective cooperative action on and with the changing natural and social environments so that practical good results. From this perspective, *techne* is not only ubiquitous in human life but is necessary to the human hopes of knowing truth and practicing good. Each kind of technology finds its appropriate end when it 'melts into' or becomes the scene of truth or good.

But because *techne* has no knowledge of its own ends, the use of rational and productive technologies must be disciplined by something else. Productive technology should be disciplined by *phronesis* and in this way it will assume its authentic place in practice. Rational technology should be disciplined by *sophia*, or wisdom about dialectic, and about the reciprocity of theory and practice. Yet here is a paradox: each technology must be disciplined by something which is untechnical, but that something can only be acquired by finite beings through involvement with some technology. This is not a vicious circle, however: *phronesis*, for example, is not caused by the productive technology it must then discipline. What is the source of *phronesis* and wisdom? Here it is enough to say that *phronesis* is not caused nor does it have a source, as though it were a commodity. To adopt Gadamer's ontological language, *phronesis* is the practical apprehension of one's participation in being: it represents what is gained between the moment I see myself as merely a finite being imprisoned in my finitude, and the moment I see myself as a finite being liberated to infinite dialogue. Of course, this is not the only way to speak about the source of *phronesis*, but it is the way presented by Gadamer's work.

What, then, of *techne* which deforms practice, to which Gadamer and many others have given attention, or of *techne* which deforms philosophy? This occurs when a technology ceases to 'enable-with' practice or ceases to 'enable-with' theory, and is substituted for the thing it should be aiding. When productive technology dominates the pursuit of good we have arrived at *technicized practice*, at practical choices which are arranged to attempt to control the appearance of good by terminating or manipulating dialogue, and by dictating which material and non-material tools represent this good. By this reasoning, in fact, 'technicized practice' is oxymoronic: it is the choice to eliminate choice, an announcement of the end of conversation. As an ideal-type, technocracy, for example, would be the political order of social practice which is built on technicized practice, in which every practice is dictated. Correspondingly, when rational technology dominates dialectic, instead of wisdom we arrive at *technicized theory*, that is, at beliefs and claims which suppress dialectic into the certification of a single, controlled, produced truth. Ideology, in its pejorative sense, is one way of speaking about the domination wrought by rational technology.

But what is the end to which technicized practice and technicized philosophy are devoted? Or since *techne* has no knowledge of its end, does this mean that there is no goal or purpose to which technicized practice and technicized philosophy are devoted? In both cases, there is an attempt to evade or to control the finitude and historicity of human life by producing and guaranteeing the good and the truth which otherwise can appear to be so illusive or impossible to us. In one sense, the ends are indeed specific ends; given the historicity of human life, it cannot be otherwise, and we can all point to personal and societal instances of such evil ends and their evil consequences. But in another sense, one which Gadamer himself suggests, there is no end when life is technicized: to choose to eliminate choice is a repudiation of practice as much as it is a practical act; to assert as a truth that which denies the dialogical character of truth is a denial of theorizing as much as it is a product of theorizing.[83]

Techne has the same relationship to evil in technicized practice as it has to good in good practice. Such technical knowledge and skill is not evil itself nor does it cause evil, yet it is present in the scene of evil and evil could not occur without its presence. Thus, *techne* can 'enable-with' both good and evil. The choice to evade or control the historicity of life finds in technical knowledge and skill the only way to express and pursue this choice; similarly, the choice to concretize good with and in human historicity finds in *techne* the only way to express and pursue this choice.

We can now apply this analysis to our initial questions. If good appears in the world, and thus in the midst of the world's technicity, we must conclude that good and *techne* are not antithetical, and append this reminder to Gadamer's critical principle that *techne* does not create good. Practice and theory always appear with corresponding forms of *techne*, yet practice and theory simultaneously go beyond those forms. With respect to practice, this means that the presence of *techne* can hardly be regarded as necessarily detrimental to practice. Nor would we want to 'purify' practice of it: in the appearance of good, *techne* must be present. The task of *phronesis* is not to utterly avoid or purify itself of technology, but to discipline the desire to use *techne* to determine practice, to struggle to use it as wisely as one can.

Similarly, a form of *techne* is present in practical philosophy and should not be regarded as a foreign body which must be expelled. *Techne* is present, for example, in Gadamer's recourse to 'rule and case' language to describe concretization even though concretization is finally quite different from that mode of thinking. When practical philosophy participates in practical deliberation, one must recognize the technical character of such language if one is to let go of it and move beyond it to genuine concretization. The abstract or metaphorical targets which practical philosophy provides to *phronesis* show both the use and limits of rational technology: those targets point our attention in the right direction but cannot identify for us what is the final and real target of good practice. Consequently, we should not allow the kinds of *techne* appropriate to theory and practice to collapse into a single 'super-*techne*.' Of course, the deformation of *phronesis* and practical philosophy is based on a viola-

---

[83] See "Discussion on 'Hermeneutics and Social Science,'" 332: "I do not know what it means to say that technological society has an end. It does not have an end. It is characterized precisely by the fact that what can be produced has to be produced."

tion of these 'rules.' The fact that rational technology and productive technology perform analogous functions for theory and practice, respectively, can be easily used as an excuse for postulating a supreme technology by which to provide certain truth and to control the appearance of good. This would constitute the deformation which is specific to the unique cases of practical philosophy and *phronesis*. Not only has theory here been dominated by rational technology, and practice dominated by productive technology, but the distinction between these realms of activity has been erased.

The net result of these preliminary reflections is that a hermeneutical philosophy of practice should present *techne* as essential to the scene in which good appears, yet not the source of good. Chapter Three elucidated how Gadamer develops a view of the relationship between word and 'world,' and this relationship appears to be analogous to the relationship between *techne* and practical good: as word "exists only in order to disappear into what is said,"[84] so *techne* could be said to 'disappear' into good.[85]

ON THE DANGER OF THE PHILOSOPHER'S SOCIAL ARROGANCE

The cautionary orientation of Gadamer's work with regard to the two dangers we have been discussing is reinforced by his concept of the philosopher's social role, or more accurately, by his restriction of that role. His attention to this topic is limited, and is hardly developed in a systematic way. But his general attitude is represented by this polemical statement: "However much he [the philosopher] may be called to draw radical inferences from everything, the role of prophet, of Cassandra, of preacher or of know-it-all does not suit him."[86] Elsewhere, in a public discussion, he says, "I think that the philosopher becomes a ridiculous figure when he claims to do more than correct and criticize philoso-

---

84 *TM*, 475 [422].

85 Whether this is more than an analogy and is perhaps an affinity or commonality between the relationships of 'language-truth' and '*techne*-good' is an intriguing idea, but must be left outside the scope of this investigation. It does suggest, however, that the kind of dialectic between finite and infinite which Gadamer finds in linguistic experience may also have a related appearance in another basic constituent of human historicity, that of *techne*. Such an idea does not necessarily contest the universality of language affirmed by Gadamer; it rather suggests that both 'words' and 'techniques' participate in some kind of dialectic between finite and infinite.

86 *TM*, xxxviii [xxiv].

phy."[87] In short, Gadamer restricts the role of philosopher to avoid activities he thinks inappropriate in some way. In part, this is to avoid absolutist claims, or claims about things in which the philosopher has no special competence. It is also, by implication, to discourage situations where others assume, or society in general assumes, that deference should be given to the supposedly wiser judgment of the philosopher in matters which are not really philosophical and in fact concern everyone. Given such concerns, one wonders if Gadamer might not view the philosopher who engages in the tasks of the participatory function as "a ridiculous figure," as one who makes "radical inferences" and who thinks less educated persons should defer to those who are wiser, as the Sophists once taught.

Let us be sure we correctly understand Gadamer on this matter. In his interview about Strauss, Gadamer says

> Academic teachers always come too late. In the best instance, they can train young scholars, but their function is not to build up character. After the war, I was invited to give a lecture in Frankfurt on what the German professor thinks of his role as an educator. The point that I made was that professors have no role to play in that regard. Implied in the question at hand is a certain overestimation of the possible impact of the theoretical man.[88]

It is altogether a good thing—and a healthy sign on the part of any philosopher!—not to overestimate the practical or moral influence which philosophers do have, or ought to have, in society. But it seems excessively onesided to simply assert that the philosopher has "no role," by which it seems clear he means no political role as an educator.

Speaking about moral claims and truth claims in general, Frederick Lawrence correctly observes that Gadamer leaves the impression "that it is almost wrong-headed to critically ground what bears us along." Lawrence also voices concern about a further implication of Gadamer's perspective, though he says he trusts Gadamer would not actually assent to it: "that it is not of the utmost importance to discriminate between the limits of human finitude on the one hand, and downright human evil

---

[87] "Discussion on 'Hermeneutics and Social Science,'" 322. It is interesting that Gadamer immediately goes on from this to say, ". . . and it is an inadequate philosophy that does not realize the one-sidedness of the orientation of our civilization." This is exactly the kind of statement which does more than just "correct and criticize philosophy."

[88] "Gadamer on Strauss," 10.

and irrationality on the other."[89] In the public forum where Lawrence posed these concerns, Gadamer gave his reply, after first recapitulating Lawrence's concern:

> . . . that by dispensing with the question of the *Lebenswelt* and attempting to provide a substitute by reference to Aristotle's work, I abandon the quest for a philosophical principle of discrimination between good and evil. My response is that it would be very dangerous for the philosopher to take a position of privilege over against the rest of society on this issue. I cannot believe that it is the function of the philosopher to formulate in the form of a principle that which should be done in the concrete.[90]

Of course, this response illustrates Gadamer's desire to prevent the philosopher's pronouncements from being applied in a technical fashion. But by implication it also shows his tendency to view all theoretical questions about good and evil as requests for techniques that 'solve' practical problems. The first of these intentions is prudent; the second takes more precaution than he needs, or than he could justify.

An exchange on this subject between Bernstein and Gadamer also bears on this matter. Bernstein poses a challenge to Gadamer that is not unlike our own:

> If we follow out the logic of Gadamer's own line of thinking, if we are really concerned with the "sense of what is feasible, what is possible, what is correct here and now," then this demands that we turn our attention to the question of how we can nurture the type of communities required for the flourishing of *phronesis*.[91]

In his reply to Bernstein, Gadamer says that "I too am in favor of a government and politics that would allow for mutual understanding and the freedom of all . . . But I am not talking [in his works?] about what is to be done in order to realize this state of affairs."[92] His exclusion of pragmatic concerns is expressed here only as his personal choice, and not necessarily as the consequence of a philosophical principle. But it still seems clear that he does not want his words to be viewed as the elevated pronouncement of a philosopher, and perhaps not even as the implication of his own philosophy. In any case, his reply once more indicates that he

---

[89] Frederick Lawrence, "Responses to 'Hermeneutics and Social Science,'" *Cultural Hermeneutics* 2 (1975): 323, 324.

[90] "Summation," 330.

[91] Bernstein, *Beyond Objectivism and Relativism*, 158. His quotation is from *TM*, 1st ed., [xxv], which appears at xxxviii in *TM*, 2nd ed.

[92] Appendix. A letter," 264.

tends to lump reasoning about practice either into the highest abstractions of the reflective function, which he is willing to address, or into the most concrete kind of pragmatic judgment, which he does not address, apparently without considering the possibility or utility of some other levels of reasoning beyond these stark alternatives. Gadamer goes on to make this further response to Bernstein's statement above:

> The use you make of my expressions about the concrete, the real and the feasible, as opposed to pure abstraction, is odd. It sounds as if I wanted to interfere with this task of concretization through my philosophic work. This is precisely what I do not want; I only intend to remind the theoreticians that this [i.e., concretization] is what it all hinges on in the final analysis.[93]

Gadamer apparently does not see how he has avoided, or perhaps even interfered with, the task of concretization when he takes measures of this kind to prevent his own practical philosophy from addressing the participatory function.

In one passage, perhaps the only passage of its kind, Gadamer explicitly concedes that the philosopher as philosopher cannot avoid directly addressing practical issues. However, he does not affirm this point as much as allow himself to be dragged to it, and this illustrates his general reluctance to grapple with topics associated with the participatory function of practical philosophy.

> Yet perhaps, all in all, it is a contradiction in terms at one and the same time to suppose that reason ought to have power and to practice government, and to suppose it wholly proper that the strange tribe of philosophers remains almost invisible in the contest of real power struggles between people, states, classes, religions, outlooks and economic systems.[94]

It may be that Gadamer is negatively influenced by the example of Heidegger, whose support for German National Socialism in the early 1930's "horrified" Gadamer at the time, and has raised much controversy over the years. Gadamer's assessment is that Heidegger was not politically sophisticated enough to grasp what he was getting into, and that Heidegger's intentions were not compatible with the means he accepted. Gadamer is also quite aware of the damage those actions did to the legacy of Heidegger's work. These reactions are evident in a short com-

---

[93] "Appendix. A Letter," 264.
[94] "The Power of Reason," 14.

mentary Gadamer wrote[95] upon the publication in 1988 of Victor Faria's book, *Heidegger and Nazism*.[96] Some or all of these reactions may have contributed to a decision by Gadamer that, in his own case, it would be better to steer clear of all but the most general political discussions or endorsements.

However, we are not necessarily asking the philosopher to get as politically involved as was Heidegger—or as have other philosophers in the name of other causes. Rather, we are asking whether the role of philosopher—as it applies to either the judge or the object of judgment—should have any bearing on our political practice. The evidence seems to show that Gadamer answers this question negatively. But he seems quite prepared to respond affirmatively to another and more common question: 'Should we—as human beings—make judgments about the cause someone else chooses?' Are these two questions so unrelated, or is there a relationship between one's humanity and the vocation by which one manifests that humanity, a relationship that needs attention?

A more appropriate concept of the philosopher's social role can be developed in terms of the concept of dialogue. Instead of trying to completely disengage the philosopher's role from the citizen's role, we should attend to the fact that the philosopher and the citizen are first human beings. In particular, three points can be made about how these roles of philosopher and of citizen intersect in our humanity. The first point is that dialogue is a demonstration of the virtue of judging-with, a virtue which should apply to philosophizing as much as it does to moral practice, and therefore as much to the philosopher as to the practical

---

[95]   Hans-Georg Gadamer, "'Back from Syracuse?'" trans. John McCumber, *Critical Inquiry* 15 (Winter 1989): 427-430.

[96]   Of Heidegger's actions and motives at that time, Gadamer makes this statement:

> If we wish to dignify his political engagement by calling it a political 'standpoint,' it would be far better to call it a political 'illusion,' which had notably little to do with political reality. If Heidegger later, in the face of all realities, would again dream his dream from those days, the dream of a 'people's religion,' the later version would embrace his deep disappointment over the actual course of affairs . . . Earlier, in 1933 and 1934, he thought he was following his dream, and fulfilling his deepest philosophical mission, when he tried to revolutionize the university from the ground up. It was for that that he did everything that horrified us at that time. For him the sole issue was to break the political influence of the church and the tenacity of academic bossdom. (428)

agent. To philosophize via dialogue means to concretize in practical philosophy itself the point of its relevance to practice; it means providing answers to moral questions, such as Lawrence's question about distinguishing good and evil, answers which attempt to judge-with the interlocutor, which testify to moral truth in a necessarily finite way, but nevertheless invite the discovery of an infinite common reality. This is how Gadamer's own philosophy describes every event of human understanding, and it must also apply to what the philosopher does. Gadamer's simple but insightful observation, below, into the connection between the event of dialogue and the moral value of solidarity, accurately portrays how the philosopher of practice should interact with his or her audience—but which Gadamer fails to adequately incarnate in his own philosophy:

> My point is that the dialogue is a good model for the process of overcoming the structure of two opposing postures. Finding a common language is not contributing to a new handbook of science or thought; it is sharing in a social act. This is a rather useful conclusion—to discover that the process of dialogue and all that is involved in its unfolding actually consists in an ongoing effort to bridge any form of alienation.[97]

To engage in dialogue characterized by the virtue of judging-with is not merely a passive activity of mutual interpretation. It involves simultaneously asserting one's position with confidence, and risking it by

---

[97] Hans-Georg Gadamer and Paul Ricoeur, "Conflict of Interpretations" [containing one address by each author], in *Phenomenology, Dialogues and Bridges*, ed. Ronald Bruzina and Bruce Wilshire (Albany, NY: SUNY Press, 1982), 304.

This aspect of philosophical hermeneutics has particular relevance to philosophical reflection on several areas of human life and practice. For example, in so-called comparative philosophy, horizons from different lands and cultures are themselves the (potential) participants in a dialogue: each 'knows' things unknown to others, yet to share with and learn from the others, each must reach and risk beyond itself with hard but open questions.

It seems that this may be the kind of activity which Raimundo Panikkar has in mind when he concludes that comparative philosophy, as a field, must be conceived as a "diatopical hermeneutics." He defines this as the discipline of interpretation which is appropriate when "two (or more) cultures . . . have independently developed in different spaces (*topoi*) their own methods of philosophizing and ways of reaching intelligibility along with their proper categories." Panikkar's solution to this basic contemporary problem will seem familiar to readers of the present work: "The *topoi* are connected by simply going over there and actualizing the encounter." (130, 133, in his essay "What is Comparative Philosophy Comparing?," 116-136, in *Interpreting Across Boundaries*, ed. Gerald J. Larson and Eliot Deutsch [Princeton, NJ: Princeton University Press, 1988].)

submitting it to the questioning of the other person. Assertions cannot be authentically asserted in dialogue unless they are also risked. Our second point, then, is that the claim to authority and the readiness to risk that authority must occur together in every assertion—including philosophical assertions—and thus in every act of judging-with. True dialogue will demonstrate both confident humility and humble confidence.[98]

This conceptualization of dialogue has bearing on how we should view social dialogue about practical issues and on the role of the philosopher in particular within that social dialogue. It leads to the following reflections, on which our third point will be based. Because there are practices involving everyone about which judgments must be made, and precisely because (as Gadamer reminds us) no one, including the philosopher, is wiser than all others about such judgments, everyone must assert judgments about what is common yet beyond anyone's particular competence; herein lies the meaning of social practical responsibility. Thus, the hermeneutical perspective implies a vision of society in which everyone must apply their professional or vocational knowledge competencies to all those practical questions about which no one appears 'competent' to judge, and at the same time be ready to risk interaction—in agreement and opposition, in dependence and contention—with those who arrive at a variety of different judgments, in part because of their differing competencies.[99] The necessity to step beyond our competence can, however, become the opportunity for social good to appear: it is by entering a dialogue in which we all must judge and judge-with that we discover what friendship and solidarity mean. From this perspective, the hermeneutical philosophy of practice echoes its Aristotelian ancestry: one's role as citizen—certainly of a society, but we would add of all human society, and ultimately one's role as citizen of the realm of being—is both the origin and end of all other roles.

We must therefore oppose Gadamer's implication that (to adapt his own phrase regarding the scientist) insofar as one is a philosopher, the

---

[98] The ideas developed here have roots in Gadamer's own statement, quoted in Chapter One, the implications of which he seems to have compromised in the development of his practical philosophy: "In fact our own prejudice is properly brought into play by being put at risk. Only by being given full play is it able to experience the other's claim to truth and make it possible for him to have full play himself." (*TM*, 299 [266].)

[99] Beiner points to the hermeneutical rationale for this argument when he states that, "Our ends are not merely pursued rhetorically, they are themselves constituted rhetorically. This is what it means to say that political ends are subject to deliberation (and not simply manipulation)." (Beiner, *Political Judgment*, 95.)

demand to judge and act in the public interest is inapplicable.[100] Philosophers will not fulfill either their vocational role or their role as citizens by confining theoretical and practical judgments about the social good to a scope in which each feels his or her correctness will be guaranteed by competence in a particular field of knowledge. Rather, and this is our third point, the knowledge competence of the philosopher is an integral part of his or her role as citizen; therefore, neither the philosopher nor society at large can prevent that competence from bearing on practical and moral questions, nor would we want to if community is truly our goal. The authority which the philosopher has in his or her field of knowledge is not something to be manipulated for ideological or self-serving purposes, but neither is it to be concealed; for if it is disciplined by judging-with, the philosopher's assertions of authority will prove, instead of being dogmatic, to be invitations to dialogue.

Gadamer writes: "The dialectician or the philosopher, who really is one and not a Sophist, does not possess a special knowledge, but in his person he is the embodiment of dialectics or of philosophy."[101] This statement is a better guide to the role of the practical philosopher than Gadamer seems to realize. Too often Gadamer treats practical philosophy as though his statement meant that since we lack the foundational knowledge by which to easily resolve all practical problems, philosophers should be silent about such things. Philosophers should not 'get their hands dirty' in issues of social and political practice. To us, however, this statement means the opposite: it is the knowledge of dialectic which liberates the philosopher to participate in practical judgments with confidence. If, as Gadamer testifies, we can fully embrace the historicity of human existence, then the reasoning of the philosopher—and of every citizen—can follow wherever dialogue leads. And dialogue always starts with questions of everyday practice.[102]

---

100   This seems a fair inference about Gadamer's view of the philosopher's social role, even of every vocation's social role, from the following statement about the scientist, quoted in Chapter Five: "One can demand of the scientist insofar as he is a citizen that, as citizen, he act in the public interest; however, insofar as he is a scientist this demand is inapplicable." ("Summation," 330.)

101   *RAS*, 122.

102   This brings us back to Gadamer's response to Heidegger's political choices in his short piece, "'Back from Syracuse?'" Gadamer has the good sense to acknowledge—but only implicitly—certain things about Heidegger which also illustrate some of the concrete consequences of the perspective just outlined. First, the philosopher may fail to be wise in his or her practical choices, but in itself this is no reason to deny him or her the right to try to apply lessons drawn from philosophical reflection in

Recall, finally, our discussion in Chapter Five of Gadamer's 1965 address, "Notes on Planning for the Future," given to a conference of about

---

political practice. Gadamer chastises Heidegger, not for his political activity per se, but, in effect, for the stupidity of his practice, for his lack of *phronesis*.

> That Heidegger's revolution in the universities failed, and that his involvement in the cultural politics of the Third Reich was a sad story we watched at a distance with anxiety, has led many to think about what Plato came up against in Syracuse. Indeed, after Heidegger resigned from the rectorate, one of his Freiburg friends, seeing him in the streetcar, greeted him: "Back from Syracuse?" (429)

Some may think that such a failure—by Plato as well as Heidegger—should imply a proscription against all political practice by philosophers. This is not our position, nor is it an inference conveyed here by Gadamer. As for those who embrace such a view, they must truly live in ivory towers. The failure in Freiburg was not only due to the limitations of those the philosopher hoped to edify, or to a distorted practice of philosophizing; it also lies in ignorance of the unsurprising fact that while practice may benefit from philosophical insight, it requires more than just this.

Second, the philosopher (and citizen) cannot claim that theorizing has no effect on his or her practice, that theory exists in an amoral realm separate from practice. No practice can be reduced to a theory, even to the theory that supposedly inspired it; but neither is practice unconnected to such theorizing. If we are indeed subject to the history of effects, and to the reciprocity of theory and practice, then Heidegger's bad choices, like his good ones, have antecedents which also left traces in his philosophy, no matter how elusive those connections may be. It is quite important that Gadamer essentially agrees with this, at least as this is evident in his comment on Heidegger: "It has been claimed, out of admiration for the great thinker [Heidegger], that his political errors have nothing to do with his philosophy. If only we could be content with that! Wholly unnoticed was how damaging such a 'defense' of so important a thinker really is." (428)

Third, while philosophy can influence or guide practice, no philosophy is reducible to its practical and historical effects, not even to the effects—for evil or good—for which its author is responsible. Philosophy is not only a practice; it is first intended as a testimony to things that outlast every witness. Below Gadamer seems to speak of his own debt to Heidegger in this regard; the point, however, has wider relevance:

> I get asked whether, after these revelations (which for us [Heidegger's colleagues] are no such thing), one can "still even today" have anything at all to do with the philosophy of this man. "Still even today"? Whoever asks that has much ground to cover. What was received . . . as a major spiritual renewal was Heidegger's lifelong altercation with the Greeks, with Hegel, and finally with Neitzsche. Did that all become fraudulent? Have we absorbed or gotten beyond it? Or is the real point [of such questions], perhaps, that people should not think at all, but only follow a completed ideological-political recipe or apply a system of rules worked out by social science? (429)

two dozen academics and policy-makers who were considering the "Conditions of World Order." We earlier discussed the internal, philosophical merits of his address; here the question, to put it bluntly, concerns own Gadamer's practice, and specifically his practice of philosophical reflection and speaking. Under the conditions he faced, how far did his participation in this conference effect the good he sought? Did he show the kind of *phronesis* about his potential impact on this group—a microcosm of some of the minds that shape our world—to which his philosophy so often draws our attention? There is little by which to measure this, except by reference to the report on the conference compiled by Stanley Hoffman. What does it signify that in twenty-five pages of this report, in which several widely recognized names receive much attention (which may reflect either the consequence or the cause of their notoriety), Gadamer is mentioned only a half-dozen times, very briefly and without much significance? The lack of significance is further evident when one notes that in the report's section regarding the conference's discussions on "Thought and Action," including sub-headings on values and world order, on the sciences, social sciences, and action, and on the role of intellectuals—all subjects on which Gadamer has sought to lay particular stress—his name does not appear once in any substantive context.[103]

Some of the possible reasons for this relative eclipse of our philosopher do not concern Gadamer's shortcomings, but those of his interlocutors. Hoffman's report (if it accurately reflects the course of debate) as much as suggests that on hearing Gadamer's address, the participants paused a few moments to murmur "yes, yes," and then returned as before to the babel of debating how, exactly, to plan the future. Is it the philosopher's fault if his or her message is ignored?

It is, however, the task of the *phronimos*, the practically wise person, to apply his or her competencies toward what is good in an appropriate manner; to fashion his or her dialogue and rhetoric so as to really meet them where they stand, without trying to climb above them or cower below them—and discover a new path together. Few of us are very adept at the ways of *phronesis*; but all of us know it when we see it, and all of us are charged with doing what we can to incarnate it. And when the philosophy of practice is applied in our own practices, those who reflect on such things should be better able to demonstrate their significance than Hoffman's report indicates of Gadamer. It is true that Gadamer did articulate his perennial agenda, and there are signs that he ardently sought

---

[103] Stanley Hoffman, "Report on the Conference."

to tailor the presentation of that agenda to his audience. But according to the evidence of Hoffman's report, partial or uncorroborated as it may be, Gadamer failed to have rhetorical success in this instance. Is it the audience's fault if they are not moved toward the purpose of practice by a speaker who fails to adequately judge-with them?

Gadamer's closing to his address demonstrates his desire to present his intellectual peers with something truly challenging, and at the same time the absence of a concrete engagement with any technocratic presumptions he may have encountered among his conference colleagues.[104] He tries to answer why we ought to care about whether philosophy can still function to cause the "awakening of consciousness." He rejects the pathway toward this function embodied in Heidegger or in Buber, in which philosophy is seen "as a kind of secularized eschatology," which "expects nothing definite" but stands as "a kind of challenge." If we read him correctly, he regards this as an irresponsible concession to simply letting events take their course, to waiting for time to mutate the world until it arrives at the preferred form of reality. The alternative pathway to a new consciousness (and therefore to a definition of philosophy), the pathway he endorses, is "to become aware of how little things change, even where everything appears to be changing."[105] With this unfashionable thought, and perhaps one unexpected by his conference colleagues, Gadamer no doubt intended to interject the fruit of his endeavors which he thought most profound, precisely in light of the technocratic temptation.

> It could be that the technological dream entertained by our time is really just a dream . . . which, when compared to the actual realities of life, has a phantom-like and arbitrary character . . . Continents and empires, revolutions in power and in thought, the planning and organization of life on our planet and outside it, will not be able to exceed a measure which perhaps no one knows and to which, nevertheless, all are subject.[106]

The problem with Gadamer's message is not that it is reactionary; it is not, but in fact is a corrective which the technocratic mind should particularly heed. Nor is the problem that he is not pragmatic enough or realistic enough, although more attention to such tasks is necessary and would have been welcome; the 'prophetic' function of a radical correc-

---

[104] Or alternatively, perhaps he was the one who spoke too abstractly, and some of his peers were more concrete than he.
[105] "Notes on Planning," 589.
[106] "Notes on Planning," 589.

tive such as Gadamer offers has a very realistic function, which is little understood in a world unfamiliar with being shaken by words that arrive from beyond us and have quite concrete consequences. But when deliberation toward practice is in question, these merits cannot mark the culmination of reflection, and must instead constitute the premise from which we embark. The problem is that Gadamer does not fully meet his interlocutor where he or she stands, ready to walk with him or her until both have arrived at a new understanding. We must deposit our truth at the threshold of another's responsibility, not at the border of our own competence. The implications of dialogue reach even this far—in fact they reach here immediately. This is the concrete consequence of Gadamer's own statement that "precisely and especially practical and political reason can only be realized and transmitted dialogically." And that is how we can "vindicate again the noblest task of the citizen—decision-making according to one's own responsibility instead of conceding that task to the expert."[107]

CONCLUSION

It is noteworthy that each of the three dangers about which we have seen Gadamer demonstrate concern corresponds to one of the three Aristotelian branches of knowledge: the danger of foundationalism concerns the proper expectations of theory, the danger of technologism concerns the correct attitude toward *techne*, and the danger of the philosopher's presumptuousness (or of society's deference) concerns the obligations and opportunities of the individual and his or her vocation. The measures Gadamer takes to protect practical philosophy from the temptations of arrogance in each realm certainly testify to his comprehensiveness, although this correspondence to Aristotle hardly seems to have been planned. Unfortunately, these protective measures are also evidence of a degree of myopia, born of an anxious concern to avoid these dangers. This myopia, while now more understandable, and certainly not unique to Gadamer, also distorts the task—it slants the question—which practical philosophy pursues. In fact, in his own practical philosophy, Gadamer protects the fruits of his labor so carefully that they suffocate. As well-intentioned and as humanly familiar as Gadamer's reasons are, his silence amounts to a reluctance for which there is no adequate

---

[107] "Hermeneutics and Social Science," 316.

practical need or hermeneutical justification. Indeed, experience in these realms shows that this excess of caution has no place in practical philosophy.

Another interpretation is possible of the conditions of human existence without being dishonest about the dangers it contains. The legitimate capacity of theory is a matter of articulation, of the soul's necessity to utter a worthy reply to the cosmos that calls him or her into being; the proper use of *techne* is a matter of first acknowledging that it is already woven into our life without concluding that it therefore determines our life; and the correct practice of vocation is, in the final analysis, simply a matter of exercising dialogue with one's neighbors by reaching out from—and therefore beyond—the place defined by one's competencies.

None of these notions is really foreign to Gadamer's thinking. Nor is our general critique of Gadamer's practical philosophy based on principles other than what can be found in his works. Our defense of such an application is predicated first on the purpose served by practical philosophy. But the validation of that purpose within the horizon of hermeneutical philosophizing is something which Gadamer's supporters have not pursued; consequently, it has not been demonstrated to his critics either. Nor, most importantly, was it clear that Gadamer himself fully endorsed the inference which follows from this defense: not even the dangers to which he points should lead us to disregard the realities which constitute our historicity—and simultaneously bless human life with the possibility of goodness.

Epilogue

One more question remains to be addressed here. We have concluded that we can and should proceed to articulate a more comprehensive practical philosophy than Gadamer offers. Of course, to proceed
means that inevitably, in yet unknown instances and circumstances, we
will fall to temptations, we will corrupt practice, and we will diminish,
twist, and even destroy the goodness we already have. While the potential gain of taking such risks may be admissible in theory, in practical life
the arguments we have presented are still not compelling as long as we
are inclined to think that caution may be the better course after all. What
will inspire us to abandon caution and take those risks?[1]

This is a question which Gadamer neither poses nor answers, at least
not explicitly. In fact, the answer which is almost hidden within his work
is at odds with his well-meaning but mistaken efforts to contain those
risks, and even to eliminate them. The dangers of foundationalism,
technicized practice, and arrogance toward society are certainly real

---

1 It may be asked whether this is in fact a question that practical philosophy can
address. To think so only means to grant that genuine answers to that question will
not be irrational or extra-rational. This does not require new (or old) compromises
over what reason can show, but it does perhaps require a reconception of reason.
Genuine answers to the question we have posed merely go one step beyond the
monological reason of technocracy to the reason on which dialogue depends and
which dialogue activates. It cannot be that reason is a key which opens all doors
except the very one to our deepest questions and hopes; rather, it is the soft but
unbreakable thread that leads us into every mystery as far as the human mind can
bear, and continues beyond. If we think this asks too much of reason, perhaps it is
because we have abandoned the leading of that thread.

enough, and we should exert effort to avoid them. But in Gadamer's protective measures against such dangers, we repeatedly sensed that these efforts share a common, though latent concern. He seems to worry that it is not enough to be simply warned of those dangers. We must be sure we have avoided them at all cost, in order to remain true to our finitude in all our judgments and actions; therein lies the best safeguard of good practice. At the bottom of Gadamer's reluctance to further develop a new practical philosophy lies this anxious posture—one remarkably reminiscent of the radical historicist.

Actually, this is an equal and sometimes even greater threat to practical philosophy than any of the dangers of which he warns us. Theorizing from this perspective will always retreat into the perceived safety of ever more general and ineffectual statements. Moreover, as our responses to Gadamer have suggested, measures which are applied in this anxious spirit will themselves take on the character of techniques to guarantee purity, and defeat the purpose they were intended to serve.

Of course, as we noted, these arguments are not enough to dispel an anxious caution. But our observation of that caution in Gadamer's work presents an opportunity to close this investigation by returning to the ontological theme by which we can address the question we have posed. The fulfillment of this task must be left to a future effort, but here we can pass on to that project the truth contained in philosophical hermeneutics, the truth on which a hermeneutical philosophy of practice would be built.

That truth is an oblique and subtle truth, shaped by the horizon and the vocabulary of a particular purpose, place and time. It is intimated in Gadamer's declaration that "the order of being is strong enough to put limits on all human confusion," and that "the changes" experienced by each age, which give the "threatening impression of a total dissolution of morality, take place on a still foundation."[2] The relevance of these statements for Gadamer's anxiety are contained in the ramifications of Gadamer's own dialogical ontology. The following passage is instructive in this regard, for it presents the dialogical ontology with particular relevance to the understanding of human practice and of practical philosophy. Opposing the idea that the ultimate aim of philosophy is to acquire the foundation or first principle of all knowledge, he proposes an alternative.

---

2   "Über die Möglichkeit," 191.

I would call . . . [this alternative] "participation," because it is what happens in human life . . .

"Participation" is a strange word. Its dialectic consists of the fact that participation is not taking parts, but in a way taking the whole. Everybody who participates in something does not take something away, so that others cannot have it. The opposite is true: by sharing, by our participating in the things in which we are participating, we enrich them, they do not become smaller, but larger. The whole life of tradition consists exactly in this enrichment so that life is our culture and our past: the whole inner store of our lives is always extending by participation.[3]

Gadamer refers here to two kinds of participation: he says we enrich life by "participating in the things in which we are participating." On the one hand we are already participants in the world in ways of which we are only partly aware, and this reflects our finitude; on the other hand, insofar as we are aware of this and choose to actively participate *in* this participation, the full meaning of our finitude appears in our capacity to reveal, share and enlarge what is common.

The ontological confidence implied by this conception of human existence was formulated earlier in our investigation in terms of the compatibility of historicity and good, and was made the overarching claim of our constructive interpretations of philosophical hermeneutics in several contexts. It was demonstrated in Chapter Three by the compatibility of philosophical hermeneutics itself with the assertion of moral universals, both because finitude involves making assertions which reach beyond what we know and beyond our limits, and because every such assertion is relativized by the historical dialogue in which it occurs. In Chapter Four, the compatibility of historicity and good was demonstrated in a different way, in the recognition that our linguisticality cannot be utterly distorted or transcended; consequently, in social practice no one can acquire any final advantage over others, nor be consigned to any final disadvantage relative to others, and it is this circumstance which can become an initiation into the power of dialogue, into its unavoidability and its sufficiency.

In Chapter Five, we followed the logic of this compatibility between historicity and good: awareness of the reciprocity of theory and practice led to formulation of the functions of practical philosophy. While we noted that Gadamer seems to resist the direction in which his own logic pulls, in Chapter Six we showed that the dangers of which he is keenly aware can be differently interpreted without abandoning his basic prin-

---

3 "Hermeneutics of Suspicion," 64.

ciples—in fact, the functions of practical philosophy require that we proceed with a different interpretation.

All these investigations merge on the identification of a truth which answers, in a quiet voice, the persistent question about our ultimate security. In response, then, to Gadamer's anxiety about remaining true to our finitude, it is possible to reflect back to Gadamer, and to each of us, the very insight to which our inquiry into philosophical hermeneutics has led us: language is always 'ahead' of us. Language is 'ahead' of our finitude, both in terms of our limitations and in terms of our inevitable failures to remain true to our finitude in our theorizing and practices. That is, to understand being as language from the hermeneutical perspective does more than only tell us about how the event of understanding occurs: it means that language compensates, as it were, for the violence we do to communication, and even for the fallibility of our confining yet excessive efforts at understanding—it compensates by maintaining our participation in dialogues of all kinds beyond our own ability to sustain that participation. This insight is what most distinguishes hermeneutical reflection from foundational or technological reflection. And through this insight, we find, almost unwittingly, a new way to understand why we can have confidence.

The dialogical ontology says one thing in two ways. In the first place, it tells us that to be finite beings means to be finite in every respect; consequently, we cannot achieve absolute control of our claims or practices in our desire to make sure they never become arrogant or technicized. Not only is it impossible to guarantee that our efforts will avoid these dangers, we are also continually involved in some manner and some measure in losing ourselves in precisely those dangers, and it is beyond our abilities to absolutely disentangle authentic assertions and practices from arrogant ones. In fact, to be preoccupied with this question is itself a foundationalist concern. But beyond this truth, the dialogical ontology also reminds us that, in effect, we are not so powerful as to be able to completely corrupt our theorizing or our practices. We fear, as well as wish, that the fate of truth and good lies in our hands, but in fact it does not: the acts by which we attempt to reach beyond our finitude are always exceeded by the power of being in the activity of language, which plays a game larger than any we can devise.

There is, it seems, a latent tension within Gadamer's work. On the one hand, it articulates a dialogical ontology which we interpret as an account of confidence in the possibility of good practice which is particularly relevant to our age of historical consciousness. Such an ontology cannot be considered only an abstract report on the nature of under-standing, but at a deeper level is also a moral witness to the meaning of the world, and to how human practice plays within it. That Gadamer is able to illuminate, even only obliquely, that solidarity is the ontological precondition for dialogue, and not only its effect; that he can remind us that solidarity is not something which we can create or crush, just by ourselves—this is of enduring value. Moreover, today, when such con-victions are hard to find anywhere, it has particular significance. On the other hand, Gadamer himself constrains his practical philosophy from relying on that witness due to an anxiety which characterizes the same 'scientific' historicism that philosophical hermeneutics is designed to cor-rect.

This tension, however, can be constructively addressed by essentially the same conviction with which the dialogical ontology addresses Strauss and Habermas: language is as much 'ahead' of Gadamer's own philosophizing as it is ahead of all the traditions and practices with which he is indirectly concerned in his texts. The problem, then, is not that Gadamer's practical philosophy is aimed in the wrong direction, but that it does not travel in that direction far enough: the barrier at which Gadamer stops is not imposed by his principles but in fact is removed by them.

* * * * *

The frontispiece to *Truth and Method* is a passage from an unidenti-fied poem by Rainer Maria Rilke, a passage which effectively frames the target at which Gadamer aims:

> Catch only what you've thrown yourself, all is
> mere skill and little gain;
> but when you're suddenly the catcher of a ball
> thrown by an eternal partner
> with accurate and measured swing
> towards you, to your center, in an arch
> from the great bridgebuilding of God;
> why catching then becomes a power—
> not yours, a world's.

Among all our efforts to participate in the life of the world, even practical philosophy can be like a ball we catch from "an eternal partner." We generally live by juggling our arrogant certainties and our perennial anxieties; the ball from beyond us interrupts our concentration, and shows most of these preoccupations to be arrogant anxieties about perennial certainties.

Yet, as an image of the relevance of philosophical hermeneutics for practice, Gadamer's selection from Rilke seems incomplete. We need something like a second verse to express—to recognize—what happens after we realize we were the target of a ball thrown by an eternal partner: we yearn to do something with the power which that ball gives to us. We want to throw it back, or throw it to someone nearby, and thus begin a game. In this lies the answer to our final question about whether the good at which practice aims is worth the risks which it always entails.

Actually, there is more to the poem than Gadamer put on his frontispiece. And this 'more' is significant because it parallels the thoughts developed in our inquiry, and perhaps also for the fact that Gadamer did not include it in his.

> . . . And when you have fully gathered
> the strength and courage to throw back,
> no, even more wonderful: when strength and courage are forgotten
> and you already *have* thrown.....(as the year
> throws the birds, the migrating flocks,
> that fling a youthful warmth to an elder across the sea—) only
> with this risk is your play made valid.
> The throw eases you no more; it hinders you
> no more. From your hands the meteor proceeds
> and rests in its open spaces...[4]

---

[4] Translated by Amy Brougher and this writer. Gadamer does not cite his source, but the original German version—consisting of both parts translated here—can be found in the last stanza of a long poem entitled, "Aus dem Nachlass des Grafen C. W." ("From the Papers of Count C. W."), in Rainer Maria Rilke, *Sämtliche Werke* (Wiesbaden: Insel-Verlag, 1957), 2: 132. The complete final stanza in the German original is as follows:

> Solang du Selbstgeworfnes fängst, ist alles
> Geschicklichkeit und läßlicher Gewinn—;
> erst wenn du plötzlich Fänger wirst des Balles,
> den eine ewige Mit-Spielerin
> dir zuwarf, deiner Mitte, in genau
> gekonntem Schwung, in einem jener Bögen
> aus Gottes großem Brücken-Bau:
> erst dann ist Fangen-Können ein Vermögen,—
> nicht deines, einer Welt. Und wenn du gar
> zurückwerfen Kraft und Mut besäßest,
> nein, wunderbarer: Mut und Kraft vergäßest
> und schon geworfen *hättest*.....(wie das Jahr

Another vision, another voice joins this conversation. It originates in a different time and place, in Moses' discourse on the covenant between Yahweh and the Israelites. It speaks of a tradition, and *for* that tradition, one to which some will look as they face the tasks ahead:

> "Yahweh your God will circumcise your heart and the heart of your descendants, until you love Yahweh your God with all your heart and soul, and so have life . . .
>
> "For this law that I enjoin on you today is not beyond your strength or beyond your reach. It is not in heaven, so that you need to wonder, 'Who will go up to heaven for us and bring it down to us, so that we may hear it?' Nor is it beyond the seas, so that you need to wonder, 'Who will cross the seas for us and bring it back to us, so that we may hear it and keep it?' No, the Word is very near to you, it is in your mouth and in your heart for your observance."[5]

Moses intended to reveal a new possibility in human existence, predicated on the concrete condition of living in covenant with God. What Moses invites is, in effect, a 'throwing back.' But the path of our inquiry suggests added reason for thinking that this possibility is unconditional—that it is received beyond that historical community and across the centuries, "with accurate and measured swing," at our center.

This covenant with God is like our participation in the infinite game identified by the ontology of philosophical hermeneutics, the game in which language is 'ahead' of us, well-prepared with inexhaustible surprises and unmovable boundaries. Just as the covenant is one in which we will not be disowned, so the game is one in which we cannot lose or be 'out.' The word we are given in both is that we do not need to seek for the ball, nor, believing we have found it, do we need to hold onto it. In fact, as Moses' discourse makes clear, the opposite is announced to be the case: the laws—and all moral imperatives and human striving—are not at all "beyond your strength or beyond your reach," but much closer than we thought. Moreover, and most contrary to expectation, we are told God means to circumcise our hearts until we

---

die Vögel wirft, die Wandervogelschwärme,
die eine ältre einer jungen Wärme
hinüberschleudert über Meere—) erst
in diesem Wagnis spielst du gültig mit.
Erleichterst dir den Wurf nicht mehr; erschwerst
dir ihn nicht mehr. Aus deinen Händen tritt
das Meteor und rast in seine Räume...

5 Deuteronomy 30:6, 11-14. *The Jerusalem Bible.*

have finished reaching and mastered throwing, until the promised communion is complete. Thus the mark of covenantal community is not meant to remind us of the burden of self-justification, but rather of God's intent and our capacity.

Works Cited

WORKS BY HANS-GEORG GADAMER

"Appendix: A Letter by Professor Hans-Georg Gadamer." Translated by James Bohman. In Richard Bernstein, *Beyond Objectivism and Relativism: Science, Hermeneutics, and Praxis*, 261-265. Philadelphia: University of Pennsylvania Press, 1983. [Includes original German text of his letter, dated June 1, 1982.]

"'Back from Syracuse?'" Trans. John McCumber. *Critical Inquiry* 15 (Winter 1989): 427-430.

"The Conflict of Interpretations." In *Phenomenology, Dialogues and Bridges*, edited by Ronald Bruzina and Bruce Wilshire, 299-304. Albany: SUNY Press, 1982. [Under this title and comprising pp. 299-320, are brief presentations by Gadamer and by Paul Ricoeur, and transcript of the group discussion which followed.]

"Correspondence concerning *Wahrheit und Methode*." Translated by George Elliot Tucker. *Independent Journal of Philosophy* 2 (1978): 5-12. [Two letters in English by Leo Strauss, and one letter by Gadamer in German with a translation, exchanged in 1961.]

Dialogue and Dialectic: Eight Hermeneutical Studies on Plato. Translated with an introduction by P. Christopher Smith. New Haven, CT: Yale University Press, 1980.

"Discussion on 'Hermeneutics and Social Science.'" *Cultural Hermeneutics* [now: *Philosophy and Social Criticism*] 2 (1975): 331-336. [Transcript of a discussion with Gadamer after his address.]

289

"Gadamer on Strauss: An Interview." Conducted by Ernest Fortin. *Interpretation* 12, no. 1 (January 1984): 1-13.

Hegel's Dialectic: Five Hermeneutical Studies. Translated with an introduction by P. Christopher Smith. New Haven, CT: Yale University Press, 1976.

"Hermeneutics and Social Science." *Cultural Hermeneutics* [now: *Philosophy and Social Criticism*] 2 (1975): 307-316.

"The Hermeneutics of Suspicion." In *Hermeneutics: Questions and Prospects*, edited by Gary Shapiro and Alan Sica, 54-65. Amherst, MA: University of Massachusetts Press, 1984.

"History of Science and Practical Philosophy." Translated by David J. Marshall, Jr. *Contemporary German Philosophy*, vol. 3 (1983):307-313. [Book review of: J. Mittelstrass. *Neuzeit und Aufklärung. Studien zur Entstebung der neuzeitlichen Wissenschaft und Philosophie*. Berlin/New York: Walter de Gruyter, 1970.]

The Idea of the Good in Platonic-Aristotelian Philosophy. Translated with an introduction and annotation by P. Christopher Smith. New Haven, CT: Yale University Press, 1985.

"Notes on Planning for the Future." *Daedalus* 95 (Spring 1966): 572-589.

"On the Scope and Function of Hermeneutical Reflection." Translated by G. B. Hess and R. E. Palmer, with additions by Gadamer. *Continuum* 8 (1970): 77-95.

"Das Ontologische Problem des Wertes." In Kuypers, K., ed. *Human Sciences and the Problem of Values*, 17-31. Hague: M. Nijhoff, 1972.

Philosophical Apprenticeships. Translated by Robert R. Sullivan. Cambridge, MA: MIT Press, 1985.

Philosophical Hermeneutics. Translated and edited by David E. Linge. Berkeley: University of California Press, 1976.

"Practical Philosophy as a Model of the Human Sciences." Translated by James Risser. *Research in Phenomenology* 9 (1979): 75-85.

"The Problem of Historical Consciousness." Text and 1975 Introduction translated by Jeff L. Close. *Graduate Faculty Philosophy Journal* 5 (Fall 1975): 1-52. (Reprinted in *Interpretive Social Science: A Reader*, pp. 103-160. Edited by Paul Rabinow and William Sullivan. Berkeley: University of California Press, 1979.) [1957 lectures first published in 1963.]

"The Power of Reason." Translated by Henry W. Johnstone, Jr. *Man and World* 3, no.1 (1970): 5-15.

Reason in the Age of Science. Translated by Frederick G. Lawrence. Cambridge, MA: MIT Press, 1981.

"Replik." In Karl-Otto Apel, et al. *Hermeneutik und Ideologiekritik,* 283-317. Frankfurt: Suhrkamp, 1971.

"Science and the Public." *Universitas: A Quarterly Review of the Arts and Sciences* 23, no. 3 (1981): 161-168.

"Summation." *Cultural Hermeneutics* [now: *Philosophy and Social Criticism*] 2 (1975): 329-330. [Gadamer's reply after hearing responses to his address, "Hermeneutics and Social Science."]

"Theory, Technology, Practice: The Task of the Science of Man." Translated by Howard Brotz. *Social Research* 44 (Autumn 1977): 529-561.

Truth and Method. Edited by Garrett Barden and John Cumming. Translated from 2nd (1965) German edition. New York: Seabury Press, 1975. [Translator unnamed; identified as W. Glen-Doepel in 2nd revised edition, xix. Includes "Hermeneutics and Historicism" (1965 version).]

Truth and Method. 2nd revised edition. Translation revised by Joel Weinsheimer and Donald G. Marshall, based on the 5th (1986) German edition. New York: Crossroad Press, 1989. [Includes "Hermeneutics and Historicism" (1965 version), and "Afterword," 3rd (1975) German edition.]

"Über der Planung der Zukunft." *Kleine Schrifte.* 4 vols. Tübingen, Germany: J. C. B. Mohr, 1967. 1: 161-178.

"Über die Möglichkeit einer philosophischen Ethik." *Kleine Schriften.* 2nd (1976) ed. 4 vols. Tübingen, Germany: J. C. B. Mohr, 1967. 1: 179-191. [Originally 1963.]

OTHER WORKS

Beiner, Ronald. *Political Judgment*. Chicago: University of Chicago Press, 1983.

_____. "Do We Need a Philosophical Ethics? Theory, Prudence, and the Primacy of *Ethos*." *The Philosophical Forum* 20, no. 3 (Spring 1989): 230-243.

Bernstein, Richard. *Beyond Objectivism and Relativism: Science, Hermeneutics, and Praxis*. Philadelphia: University of Pennsylvania Press, 1983.

_____. "Hermeneutics and its Anxieties." In *Hermeneutics and the Tradition*, ed. Daniel Bahltron, 58-70. Proceedings of the American Catholic Philosophical Association, vol. 62. Washington, DC: American Catholic Philosophical Association, 1988.

Betti, Emilio. "Hermeneutics as the General Methodology of the Geisteswissenschaften." Translated by Joseph Bleicher. In Bleicher, *Contemporary Hermeneutics: Hermeneutics as Method, Philosophy and Critique*, 51-74. London: Routledge and Kegan Paul, 1980.

Da Re, Antonio. *L'ermeneutica di Gadamer e la Filosofia Pratica*. Studi di filosofia storia della filosofia, vol.5. Rimini, Italy: Maggioli, 1982.

Habermas, Jürgen. *Communication and the Evolution of Society*. Translated by Thomas McCarthy. Boston: Beacon Press, 1979.

_____. "The Hermeneutic Claim to Universality." Translated by Joseph Bleicher. In Bleicher, *Contemporary Hermeneutics: Hermeneutics as Method, Philosophy and Critique*, 181-211. London: Routledge and Kegan Paul, 1980.

_____. *Knowledge and Human Interests*. Translated by Jeremy J. Shapiro. Boston: Beacon Press, 1971.

_____. "A Review of Gadamer's *Truth and Method*." [Translator unidentified.] In *Understanding and Social Inquiry*, 335-363. Edited by Fred Dallmayr and Thomas McCarthy. Notre Dame, IN: University of Notre Dame Press, 1977.

_____. "Summation and Response." Translated by Martha Matesich. *Continuum* 8 (1970): 205-218.

_____. "On Systematically Distorted Communication." *Inquiry* 13 (1970): 205-218.

_____. *Theory and Praxis*. Translated by John Viertel. Boston: Beacon Press, 1973.

_____. "Towards a Theory of Communicative Competence." *Inquiry* 13 (1970): 360-375.

_____. "A Postscript to Knowledge and Human Interests." Translated by Christian Lenhardt. *Philosophy of the Social Sciences* 3 (1973): 157-189.

_____. "Zu Gadamer's *Wahrheit und Methode*." In Karl-Otto Apel, et al. *Hermeneutik und Ideologiekritik*, 45-56. Frankfurt: Suhrkamp, 1971.

Heidegger, Martin. *Being and Time*. Translated by John Macquarrie and Edward Robinson. New York: Harper and Row, 1962.

Hinman, Lawrence. "Quid Facti or Quid Juris? The Fundamental Ambiguity of Gadamer's Understanding of Hermeneutics." *Philosophy and Phenomenological Research* 40 (June 1980): 512-535.

Hirsch, Eric. *Validity in Interpretation*. New Haven, CT: Yale University Press, 1967.

Hoffman, Stanley. "Report of the Conference on Conditions of World Order — June 12-19, 1965, Villa Serbelloni, Bellagio, Italy." *Daedalus* 95 (Spring 1966): 455-478.

Hoy, David. *The Critical Circle: Literature, History and Philosophical Hermeneutics*. Berkeley: University of California Press, 1978.

Humboldt, Wilhelm von. *Linguistic Variability and Intellectual Development*. Trans. George C. Buck and Frithjof A. Raven. Coral Gables: University of Miami Press, 1971. "First published 1836." Sec. 13. Passage quoted and so cited by Gadamer, *Truth and Method*, 2nd ed., 440 [*Truth and Method*, 1st edition, 399, note 70, cites *Über die Verschiedenheit des menschlichen Sprachbaus*].

Ingram, David. "Hermeneutics and Truth." In *Hermeneutics and Praxis*, edited by Robert Hollinger, 32-53. Notre Dame, IN: University of Notre Dame Press, 1985.

Lawrence, Frederick G. "Introduction." In Gadamer, *Reason in the Age of Science*, pp. ix-xxx. Cambridge, MA: MIT Press, 1981.

Lawrence, Frederick G. "Responses to 'Hermeneutics and Social Science.'" *Cultural Hermeneutics* [now: *Philosophy and Social Criticism*] 2 (1975): 321-325.

Lobkowicz, Nicholas. *Theory and Praxis: History of a Concept from Aristotle to Marx*. Notre Dame, IN: University of Notre Dame Press, 1967.

Palmer, Richard. *Hermeneutics: Interpretation Theory in Schleiermacher, Dilthey, Heidegger, and Gadamer*. Evanston, IL: Northwestern University Press, 1969.

Panikkar, Raimundo. "What is Comparative Philosophy Comparing?" In *Interpreting Across Boundaries*, edited by Gerald J. Larson and Eliot Deutsch, 116-136. Princeton, NJ: Princeton University Press, 1988.

Ricoeur, Paul. "Ethics and Culture: Habermas and Gadamer in Dialogue." In Ricoeur, *Political and Social Essays*, 243-270. Athens, OH: Ohio University Press, 1974.

_____. *Hermeneutics and the Social Sciences*. Translated and edited by John B. Thompson. Cambridge, UK: Cambridge University Press; Paris: Editions de la Maison des Sciences de l'Homme, 1981.

_____. "History and Hermeneutics." Translated by David Pellauer. *Journal of Philosophy* 73 (November 1976): 683-695.

Rilke, Rainer Maria. "Aus dem Nachlass des Grafen C. W." In *Sämtliche Werke*, pub. by Rilke-Archiv in association with Ruth Sieber-Rilke, ed. Ernst Zinn. 2: 112-132. Wiesbaden: Insel-Verlag, 1957.

Schuchman, Paul. "Aristotle's Phronesis and Gadamer's Hermeneutics." *Philosophy Today* 23 (1979): 41-50.

Strauss, Leo. "Philosophy as Rigorous Science and Political Philosophy." *Interpretation* 2, no.1 (1971): 1-9.

_____. *Political Philosophy: Six Essays by Leo Strauss*. Edited with an introduction by Hilail Gildin. Indianapolis, IN: Bobbs-Merrill, 1975.

_____. "Relativism." In *Relativism and the Study of Man*, 135-157. Edited by Helmut Schoeck. Princeton, NJ: Van Nostrand, 1961.

_____. "An Unspoken Prologue to a Public Lecture at St. John's." *Interpretation* 7, no.3 (1978): 1-3.

_____. *What is Political Philosophy*? New York: Free Press, 1959.

Strauss, Leo and Gadamer, Hans-Georg. "Correspondence Concerning *Wahrheit und Methode*." *Independent Journal of Philosophy* 2 (1978): 5-12.

Sullivan, Robert R. *Political Hermeneutics: The Early Thinking of Hans-Georg Gadamer*. University Park, PA: Pennsylvania State University Press, 1989.

Weinsheimer, Joel. *Gadamer's Hermeneutics: A Reading of Truth and Method*. New Haven, CT: Yale University Press, 1985.

Wiggins, D. R. P. "Deliberation and Practical Reason." In *Practical Reasoning*, edited by Joseph Raz, 144-152. Oxford, UK: Oxford University Press, 1978.